Windmills
A New History

Colin Moore

First published 2010

The History Press
The Mill, Brimscombe Port
Stroud, Gloucestershire, GL5 2QG
www.thehistorypress.co.uk

British Library Cataloguing in Publication Data.
A catalogue record for this book is available from the British Library.

ISBN 978 0 7524 5400 9

Typesetting and origination by The History Press
Printed in Great Britain

Contents

About the Author

COLIN MOORE graduated from Sheffield University in 1957 with an honours degree in metallurgy (BMet), and worked in the nuclear- and coal-fired power industries before joining the steel industry where he gained a Master of Metallurgy (MMet). He continued his career until retirement in the atmospheric gases sector, which serves most industries. During this time he became a Chartered Engineer (CEng), Fellow of the Institute of Metallurgists (FIM), and a Fellow of the Institute of British Foundrymen (FIBF). He also wrote three technical books, published and/or presented forty-three technical papers, and was granted fifteen patents. On retirement, he expanded his interest in restoring longcase and wall clocks and began an in-depth study of the history and engineering of windmills. To enhance such studies he joined The Lincolnshire Mills Group, The International Molinological Society and the volunteers at Ellis Mill in Lincoln, to help in showing this well-restored mill to the public and to learn more about the ancient skill of milling flour by wind-powered stones.

Windmills

In a steady wind, white sails are turning
Set above a coal-black tower.
Against a blue sky gently gleaming
As mill stones turn to make the flour.

Horse-drawn carts bring the grain by sack
Soon unloaded to the busy mill.
Rope hoist to top – not the miller's back,
The feed bins for the stones to fill.

Golden grains through chutes run down
Past the damsell, rhynd and mace
Into the stone eye to be ground
And out pours meal at what a pace.

Warm, fine, a lovely smell and brown,
To be cleaned of pollards, bran and dust
To make our daily bread in town
To old English taste that all can trust.

Colin Moore

CHAPTER 1

Introduction

This book is about windmills – defined as any device with rotating sails – their beginnings, the changes to them throughout history, and their continued evolution and growth today. The wind, however, was not the first source of power that was used by man (other than himself) to grind flour. Horses and cows were used to turn larger stones and watermills, which were in operation for over 1,000 years before windmills appeared. The actual milling process was thus already fully developed, and so it was the need for an alternative source of power – where water was not available – that triggered the invention of the first windmills.

This book looks at the innovations made by the early builders of windmills, working both to the limits of their skills and the materials that they had – wood, stone and reed. It follows the early scientists, who studied the behaviour of the wind and began to understand its enormous power, and the imaginative blacksmiths and millwrights, who decided that what had been accepted as the norm for centuries was not the end of, rather merely the basis for, new ideas.

A wheel turned by the wind could do more tasks than simply mill flour, and so men began to radically modify the windmill to pump water, saw wood, pound cloth, crush seed and powder minerals. In the fast-changing times of the Industrial Revolution they extended its useful life for another hundred years, to reach a peak of 12,000–15,000 units by the mid-nineteenth century. Steam from coal was born, and would grow to be the main source of power up to the mid-twentieth century, but where such power could not reach to remote and developing areas, the multi-bladed farm windmill evolved to meet the pressing need for water pumping, and at their zenith in the early 1930s about 6 million were in use.

There are now only a few hundred traditional windmills still about, along with a million or so farm windmills pumping water or generating electricity for farmsteads in the more remote regions of the world. The big – and future growth – is of the modern power windmill or turbine. Various forecasts estimate some 50,000 such turbines by 2020, with an installed capacity of 250,000MW. So no, the windmill did not die, merely it continued its evolution to meet current needs, as it has done since its inception over 1,000 years ago.

The key factor for the use of wind is that it is inexhaustible. It has blown since time began, and no matter how much it is used it will not run out! Man's use of wood lead to deforestation in Europe in the Middle Ages, which continues even today. The Industrial Revolution in the eighteenth century was based on coal, whose sources are being depleted. Oil and natural gas took over as the prime fuels in the twentieth century, and supplies will be exhausted certainly by the end of this century. The newest power – nuclear – may well meet future needs, but at a high cost caused by the safe processing and disposal of the waste. The wind will continue to blow till the end of time, and windmills will evolve further to use it, responding to changes in industry, trade and the political climate. The advances are always brought about by individuals, and it is often only their persistence and self-belief which makes technological changes happen. There are several such people who have been involved in the evolution and revolution of the windmill to be read about in the following chapters.

Any advancement is based on a thorough understanding of the current technology, itself dependent on measurements and data. The most important parameter for windmills is wind speed and the resultant sail speed. Up to the eighteenth century there were, of course, few instruments that could measure this, but Smeaton and Coulomb, scientists of that time, had clocks and yard sticks. Sail speed could be determined by counting the 'swish' of each sail as it passed for a minute – the Dutch called this 'enden counting' – then dividing by the number of sails, giving the sail speed in revolutions per minute (rpm). To get wind speed they released feathers and checked the time taken for the wind to blow them over a measured distance near the mill, calculating the wind speed in feet per minute.

It is also possible to estimate the wind speed from its effects on trees, flags, debris and the surfaces of streams, rivers and lakes. In 1805, Admiral Sir Francis Beaufort published his observations on the effects of wind for sailing ships, and it soon became recognised and accepted on land. Part of the Beaufort Scale is reproduced on the next page.

There are many other aspects to be noticed and wondered about. At their peak, there were some 12,000–15,000 windmills in this country, built in the corn-growing areas where watermills were not an option. Such places included the south coast from the Isle of Wight to Kent, eastern counties from Kent to Durham, the Fylde coast of Lancashire, Anglesey, Wirral and Somerset.

In round figures, one post mill with one pair of stones would meet the needs of some fifty families. The bigger tower mills, with four pairs of stones, would serve villages of 250–300 families, so the mills were built for the local communities and were paid for by the local landlord or church, who employed local carpenters,

Beaufort Wind Force

Force	Observed Effect	Wind Description	Wind Speed (mph)	Windmill
1	Flags begin to move	Light	1–3	Sails still
2	Flags unfurl; leaves rustle	Light breeze	4–7	Sails start
3	Flags extend; twigs move	Gentle breeze	8–12	Mill will run on 1, 2 stones
4	Flags flap; branches move; debris moves	Moderate breeze	13–18	Mill will run on 3, 4 stones
5	Small trees sway; wavelets form	Fresh breeze	19–24	Mill fully loaded
6	Large branches move	Strong breeze	25–31	Sails reefed to ¾
7	Whole trees sway	Moderate gale	32–38	Sails reefed to ½
8	Branches break; leaves stripped	Fresh gale	39–46	Mill stopped

blacksmiths and a millwright (if there was one). Up to the eighteenth century, the work was controlled by words and notes in a personal book, not by drawings, so the mills themselves are very individual. It is probable that, even with 15,000 windmills, no two were exactly alike, and it is this aspect that makes them truly intriguing to study.

Despite being individual, mills were not immune to political change. Though technological advances occurred, these were often in distant places and were frequently not known about, and did not seem relevant to the wind millers even when they were. The beginning of steam power from coal in the early eighteenth century was a major event in this country, and the development of roller milling in Eastern Europe in the nineteenth century was another. To study their effects on the windmills, I have relied mainly on papers, patents and books from the period when the windmills were operating, and the writers were reflecting the feelings and opinions of those times (see the literature list). Even so, those publications of the eighteenth and nineteenth century were using source material that was 400–500 years old, making the beginnings of the windmill rather difficult to define and interpret. An interesting tale is retold many times, becoming embellished and self-corroborating in the process, and finally an established fact. The actual dull truth dies through neglect, so allowing a diversity of opinion as to how things changed.

To make things more intriguing, the older smaller post mills were regarded as moveable. And, if they were named after the site, and if one mill disappeared and another had been built, or if a smock or tower mill replaced the old post mill, there can be references to both as if they co-existed. Furthermore, the mills were often named after the miller. If his sons took over, the name continued, but if the mill were sold it could have a different name after the new miller. Within the mill, because of local variations, different names were used for the same item, which makes mill studies even more fascinating. I have tried to include most variations in the following chapters, also listing them in Mill Terminology (see page 171). But who knows, there will doubtless be more – and with almost every statement made, it could be followed by except, but or however!

To aid in keeping track of one's observations, in the Appendices is a Data Sheet (see page 183) that shows the parameters required to describe a windmill in technical terms. It can be copied and filled in with as much detail as one is able – or wants – to collect. Some mills publish a lot of this information, some can be deduced from photographs in books or those taken oneself. Such pictures can be studied to gain an understanding of a piece of machinery, to see if it works as one thought on a next visit to a mill. From the data collected, one can work out – with the help of the examples in the Appendices – sail area; swept circle area; efficiency and other interesting figures, to compare the performance of mills and become another dedicated molinologist.

CHAPTER 2

The History of Milling and Windmills

Flour production from corn is the oldest industrial process in the world. In prehistoric times it changed man from a nomadic hunter to a resident farmer, when crushing corn in hollows in rocks first produced flour. The stones became pounders and the hollows became mortars. Then, saddle stones and rubbers emerged as a gentler and more consistent process. These devices are recorded in the writings of Ancient Greece and Egypt, and found as relics all over Europe and Asia. Pounding, crushing and rubbing lasted thousands of years, and until recently could still be found in remote undeveloped regions of the world.

200 BC Somewhere in the Roman Empire the quern was invented. The quern is a two-stone system, with the upper stone revolving above the base stone, producing flour by true grinding or milling. The upper stone was turned by hand, with grain fed into the central hole, and the flour and husks falling from the gap to be collected from the floor (Figure 1). The early stones would be smooth and small, say 12in–15in (300mm–375mm) in diameter. They could be rotated at about 40rpm to make a coarse – but welcome – flour, and were still used by crofters in Scotland up to 1800 when a quern cost 14s, a considerable sum then, so that several families would jointly own it. Two people would need 4 hours to grind a bushel of corn in order to make about 40lb (18kg) of flour.

100 BC The use of querns spread throughout the Roman Empire and beyond, and it was soon recognised that rotation lent itself to more than just hand power. Horizontal-wheeled watermills as early as 85 BC were used to drive larger querns, as were slaves, horses and cattle.

AD 200   Quern diameter had increased to 20in (500mm), and the upper stone had become concave, with furrows cut to promote cutting of the grains and the outward flow of the meal. These furrows or grooves, radial in their simplest form, divided the surface into segments with lands between. To begin with, the number of furrows could have been as few as four, giving rise to the name quarters. The number of main furrows rose and the increased number of segments became known as 'harps', into which secondary furrows further divided them with lands between.

AD 500   With the fall of the Roman Empire, slave mills were replaced by beast and improved vertical-wheeled watermills, with the domestic quern retained in most rural areas all over Asia, continental Europe and Britain. The majority of the mills would be powered by water and were probably using stones ranging from 3ft 6in–6ft (1.1m–1.8m), with some pattern of furrows and lands.

**AD 700
to 1100**   In the Afgan–Iran region of Asia, where in the high plateau of Seistan, Baluchistan, there are more winds than streams or rivers, someone realised that these winds could be used to turn the stones to make flour, and so the windmill was born. These windmills had a vertical shaft, directly driving a single large runner stone, 5ft–7ft (1.5m–2.1m) in diameter, so there was no need for any form of gears (Figure 2). The stone towers were built in lines, with connecting walls between to force the wind into the entry funnel of the mills, but only the sails on the funnel side could generate power. Woven slat or slit branch panels were used to deflect more wind into the funnel, or to partially close it when the wind strengthened, to give some degree of control on stone speed. The stones would be driven at 20–30rpm when the wind blew from the predicted direction. If the wind did not blow from the prevailing direction, there was no milling.

         Wherever mills began, millers would have known about the variability of the wind. A breeze moved dust and leaves and blew from any direction, whereas storms caused serious damage, so they knew that its energy was enormous. 2 tonnes of air per second pass through the sail area of a windmill with 9m-long sails at a wind speed of 15mph, but it would be 1,000 years before man could calculate this! The long road of the windmill's evolution, however, had begun.

AD 868   The first of many enigmas about the evolution of the windmill. A Danish historian, Jespersen, writing in 1953 and quoting a German journal of 1909, states that the first windmill was built in England at Croyland Abbey, Lincolnshire. The Benedictine Abbey was founded in AD 713 in memory of St Guthlac, when the nearest flowing water was nearly a mile away, so perhaps a watermill was not possible. Having struggled with hand querns or beast mills for many years, did the monks build a vertical fixed windmill? If they did, it was not copied, as the next entry about the Domesday Survey shows.

AD 1086 After the Norman Conquest of England in 1066, the Domesday Survey was carried out to determine the population and, more importantly, what they owned. The survey revealed 6,082 watermills, at annual rents of 3*d* to £3. There were probably a few hundred beast mills, but these do not seem to have been specifically recorded because they were regarded as domestic equipment, as were hand querns. The population was about 1.25 million, and one in three villages had a mill, though a substantial part of the corn was still ground in domestic querns. The grinding of flour was recognised as a prime source of regular income for the owners, but home-operated hand querns jointly owned by several families avoided any charges. What the Domesday Survey did establish was that there were no windmills in England at that time.

The watermills, using vertical wheels and horizontal shafts, needed one set of wooden gears to drive the mill stones, and in the few hundred years since furrows were first used, further refinements had been introduced. These refinements included the use of main profiled furrows that were offset from radial, separating the stone into between seven and eleven harps, which had secondary furrows set at an angle to the main ones. The lands between were roughened by tiny pits or scratches to improve milling. Curving the main furrows without secondary furrows was also used. These patterns, with only minor changes, continue to be used up to the present day, and will be fully described in the chapter on milling.

AD 1100 to 1200 This was a momentous century for the windmill. At the beginning, there were no windmills in England or Europe, but by the end there were more than fifty post mills in England and a few in France. But when and where was the first post mill built? Eminent historians have searched old church and manorial documents to deduce (rather than establish), where and by whom the earliest post mill was built. It is an absorbing detective story that has no precise conclusion. England had thousands of watermills and many beast mills that worked well, so why not build more of them?

The population was increasing, and ditch draining of wet lands to get more arable land for corn was being funded by gentry for royal grants of the land recovered. These lands were, of course, flat and unsuitable for watermills, and beasts and labour could be better employed on other tasks. Somewhere, an educated and practical man conceived the post mill,

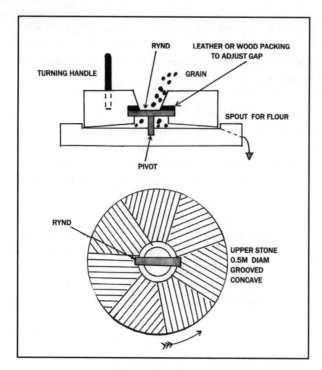

Figure 1. Hand quern about AD 200.

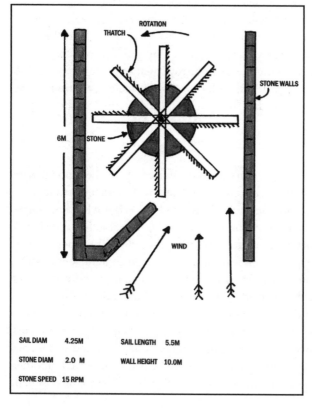

Figure 2. Vertical axis windmill, Seistan, Persia, AD 700–900.

and had the financial and manpower resources to build it. Was such a man a crusader who had seen post mills in the Middle East? There seems to be scant evidence that post mills did exist at the time of the First Crusade of 1096–1099, and in the Middle East only the vertical-shafted Siestan type existed. It is therefore most likely that it was an English invention, borne out of need for an alternative to water power. The wind is free and, in England particularly, it blows from any direction, so in using it the device had to be able to turn to face it, and thus the post mill was born.

It had a horizontal wind shaft set in a hut that could be turned into the wind. The power from the wind came from the top down – unlike the watermill, where the power came up from the waterwheel. It would be a crude structure, giving its builder a few headaches before it would work reasonably reliably and grind corn. The post could have been based on a living oak tree, whose trunk was trimmed to be the post, or on a hewn trunk buried in the ground. The sail arms were mortised through two square holes, set one behind the other in the wind shaft and wedged into it. These primitive sails would have been short, say no more than 2m long and perhaps 1m wide with 0.5m each side of the arm, and possibly initially inclined at as much as 45 degrees to the wind, but soon reduced to 20–30 degrees, which is much more efficient. The sail cloth would be stretched and tied onto a simple frame, possibly without any means of reefing, so control was achieved by turning the hut and its sails partially out the wind and at right angles to the wind to stop it. The sails swept within 0.5m of the ground, with sail tip reaching about 5m (Figure 3).

There would have been enough power to turn one set of stones of 3ft–5ft diameter, and these were driven by a big wheel on the wind shaft fitted with pegs that engaged staves in a lantern pinion arrangement – a copy of the gearing on the watermills (Figure 4). This form of drive is very tolerant to varying depth of engagement of the pegs, and misalignment of the two shafts and worn pegs or staves could be easily replaced. There may have been only one floor on the major crossbeam that supported both the stones and the hut and allowed it to turn. To hold the hut vertical, some sort of substructure with a collar round the lower part of the main post would be needed, and this sub-frame would soon become a second floor. The roof would be thatch, with wattle and daub walls. Even such a crude mill would not be cheap to build, and the expensive stones would need to be brought a reasonable distance, so the owner needed to ensure its use, as did the operators of watermills.

AD 1100 to 1134   A hint that the priory at Clerkenwell in London had a post mill.

AD 1105   A convent in France may have had a post mill.

AD 1100 to 1137   William the Almoner gifts his estate in Wigston Parva to Reading Abbey – it included a mill. Since Wigston Parva is in the centre of England – south of Leicester – with no viable water source, and it was rare to record beast mills, it is likely to have been a post mill. William had held his estate since 1100, and since the existence of the mill must predate the gift, it must have been built sometime between 1100 and 1137. In all probability, this was the first post mill.

AD 1150   The problem of grain not being taken to the mill was addressed by the establishment of Mill sokes. One of the earliest sokes was granted by Cecilia of Rumelia, Lady of the Manor of Silsden in Yorkshire, to the monks of Embsay Priory to operate the Silsden watermill. Not only were domestic querns banned, but anyone carting grain from Silsden to be ground elsewhere faced confiscation of horse, cart and grain. The soke required both tenants and freemen to take all their grain to the manorial mill, where they were charged a toll. Lady Cecilia got immediate free grinding of her grain and let the monks set their own level of toll, which was a portion of the grain ground. The system was obviously open to abuse, but because it was so profitable to the miller and to the Lord of the Manor, sokes became widespread and were rigorously enforced. Domestic querns were smashed and any resistance was overcome by force. The landlords' right to do this was upheld by law, and as a further powerful deterrent, where the mill owners were religious orders, offenders were threatened with excommunication from the Church so there would be no place in heaven for them.

AD 1155   Hugo de Plaiz gave his post mill in Iford, Sussex, to the monks of Lewes.

AD 1180   The first reliable records of windmills in Normandy and Provence, France, are made, and between 1180 and 1204 the Bishop of Chichester presented a post mill to the Church estates.

AD 1182   A post mill was recorded in Portugal.

AD 1185   The written record of the first post mill in England was at Weedly, near South Cave in East Yorkshire, which was rented for 8s per year. The actual construction of the mill is likely to predate this record. The land was owned by the Knights Templar who had fought in the crusades, lending weight to the argument that the windmill may have been introduced to England by the crusaders. However, a mill was also recorded at Camberley, built by the Bishop of Sussex, where there was no crusader connection.

AD 1191   Detailed documentation of the existence of a post mill was made because of a dispute. Dean Herbert built a post mill on his own land just outside Bury St Edmunds, where Abbot Sampson already had a total of six mills, which may or may not have included post mills. Abbot Sampson ordered the destruction of Herbert's post mill on the grounds that only *he* had the right to mill in the area governed by his abbey, whether he owned the land or not (and of course he would lose a lot of money by trade going to a rival mill). Dean Herbert protested that the wind was free to anyone, and he proposed to use the mill only for his own grain. The Abbot ordered its destruction anyway and sent his carpenters to do the job. However, the Dean had his own men remove the mill before the Abbot's gang got there, so confirming that the mills of that time were indeed small structures and could be rapidly built and – equally rapidly – taken down. The Abbot's right to do this became common law, giving the feudal mills sole right of operation, and with the mill soke, captive customers.

    In the same year, Pope Celestine III confirmed that English post mills had to pay tithes to the abbeys in whose administrative area they were built.

AD 1199   William of Wade was taken to the Royal Courts by the monks of Canterbury for building a post mill on his own land within the parish of the abbey. The mill was probably built between 1189 and 1199, and had a good established trade for grinding the locals' corn. Like Dean Herbert, he was ordered to take it down. He did not, and instead began a drawn out defence to keep his mill. In 1200, he paid the considerable sum of 10 Marks for a Royal Grant from King John to own and use the mill. Not to be outdone, the monks paid 10 Marks in 1202 to retry the case. William had to give 12 acres to the monks in compensation, but he kept the mill. In 1204, for a further 25 Marks, the court ruled in favour of the monks, and ownership of the mill went to them confirming that freemen needed Church or royal approval for mill building.

    By 1200, over fifty post mills had been built, of which twenty-three can be positively identified. No doubt even more were in the course of construction, mainly down the east coast from Northumberland and round the south coast to Sussex. The historical evidence confirms that the post mill really was an English invention.

AD 1200 to 1300   Small post mills began to be built throughout Europe for corn milling, still using short sails, and their use for drainage, even in Holland, would not be established in this period. They were still expensive, and thus only the Church or the landed gentry had the resources to fund their building and keep them running. A mill cost £10, which included the two mill stones at about £1 13s each. A labourer at this time would have only

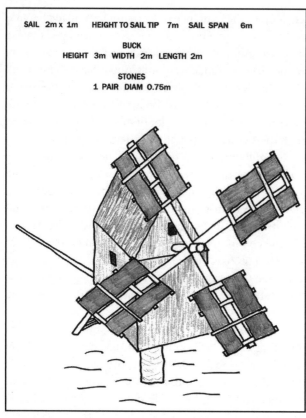

SAIL 2m x 1m   HEIGHT TO SAIL TIP 7m   SAIL SPAN   6m

BUCK
HEIGHT 3m WIDTH 2m LENGTH 2m

STONES
1 PAIR DIAM 0.75m

Figure 3. Early post mill.

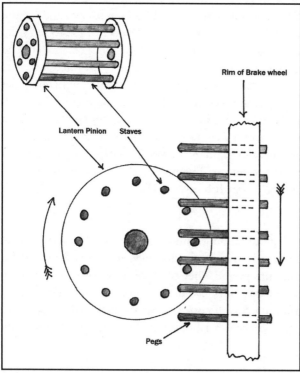

Rim of Brake wheel

Lantern Pinion   Staves

Pegs

Figure 4. Lantern Wallower.

earned about £1 5s in a year, and he would have been very happy to continue using his hand quern. So, to make sure they took their grain to the mill, more sokes were established and enforced so that all residents had to use the mill and pay its levy. Hand milling with querns was completely outlawed and the peasants could be fined for using them, or for having their grain milled outside their lord's domain.

The fine was usually 6d, almost a week's wage, so it was a very effective deterrent. Payment for milling was the toll, known also as levy, knaveship or multure, which varied from $\frac{1}{32}$ to $\frac{1}{12}$, with an average of $\frac{1}{16}$ of the flour ground. In some cases, freemen were charged slightly less than tied labour, but both had no option but to take their grain to the mill. Of the flour taken, two thirds to three quarters went to the mill owners, who also got immediate free milling of their corn, and one third to a quarter went to the miller as his wages. The miller was therefore a powerful and privileged man who was seen as being devious, cunning and to live well by deceitfully taking even more flour than the already resented toll. He was not always the honest jolly fellow that is now portrayed.

A single post or peg mill would not remain stable for long, as high winds and ground softened by rain would let the mill rock. This problem would get worse as the body was raised higher from the ground by longer posts, so props were added to help stability. Three props making a tripod mill have been recorded, but rotting of even oak would cause long-term stability problems, and in high winds the only way of stopping the mill was to turn the sails at right angles to the wind. A tilted post would make this a very hard job, and the brake wheel would not be invented until some 300 years later. At Dover Castle, they built a small stone tower mill to solve the stability problem, but, although it was the first tower mill, at this time it was not accepted, and it would be a long time before such mills became the norm.

AD 1201 King John grants a charter to the Abbot of Wellow, Grimsby, Lincolnshire, to build a post mill.

AD 1215 King John is forced by the Barons to sign the Magna Carta, granting them some self-autonomy and rights. It also established, in Article 35, that the bushel and quarter were the legal measures of the wheat trade. The bushel, however, is a volume, not a weight. It equals eight gallons, which equates to 60lb–75lb (27kg–34kg). A quarter is eight bushels, i.e. 480lb–600lb (218–272kg), and may have referred to a quarter of a horse load that was generally regarded as about 1 ton.

AD 1230 A mill soke was established at Ince, Cheshire, for all living in the area to use and for the destruction of hand querns. It was disputed, but enforced.

AD 1249   Colton Mill in Yorkshire was rented for 4s 5d per year.

AD 1286   Three men of Raistrick, Yorkshire, were fined 6d each for using hand querns in their own homes.

AD 1287   Two men of Chatteris, Cambridgeshire, were fined 6d each for 'not doing suit to the Lord's mill'.

AD 1294   John Bundeleg, the miller at Holywell-cum-Needingworth in Huntingdonshire, was fined 6d for using an oversized toll dish.

AD 1295   Dover Castle is recorded as having a parallel-sided tower mill, but it was before its time.

AD 1296   A post mill was valued at £26.

This had been a century of mill building, with windmills in new areas replacing old and small watermills and the domestic quern. It is possible that there were 10,000–12,000 mills in use in England, with more than half of them post mills, and the tower mill had made a fleeting appearance.

AD 1300
to 1700   The size of the post mill increased and became based on a manufactured oak post, to give greater height with an inclined wind shaft to improve operation. The use of wooden cogs further helped in improving the running of the mill, and this design continued in use for some 500–600 years. The mill stones were originally made from any local hard stone, but over this period it became known that Peak stones from Derbyshire were better, and worth bringing considerable distances.

The sails would be of constant pitch, and by now it would have been realised that 20–30 degrees was not the optimum. By trial and error, the pitch had progressively reduced to 15–20 degrees, with an equal width at either side of the arm. The sail cloth began to be threaded in and out of the sail bars on either side of the stocks, and the area of the sails could be adjusted by reefing. The stocks and sail bars were trimmed branches or slender trunks, referred to as medieval sails, and their length was limited to no more than 12m to be cut from a tree – 6m either side of the wind shaft – to give sails of about 5m. Initially there were no side bars, but these were added to give more strength.

At some time in this period, probably later rather than sooner, a mill builder realised that sail length could be extended considerably by fixing another spar to each side of the stock and using this to carry the sail bars and the sail. This secondary spar became known as the whip, which could be 12m long to increase the sail length to about 10m.

Subtle changes were also made to the construction of the buck to make it more weatherproof. Since the front almost always faces the wind and rain, the front boards were made to overlap the sides, and the bottom of the front was extended, like an apron, to stop the water running down the front from being blown under the buck and onto the wooden trestles.

Up to about the fourteenth century, mills had only one set of stones. Then, in the early fifteenth century, significant changes began to be made. Post mills had tail wheels added to the wind shaft to drive another pair of stones. The buck was sometimes increased to three floors, and to get even more space, porches were added to the rear and panniers to the sides. These developments exacerbated the stability problem, so more modifications to the support system were needed. Stone and then brick piers lifted the wooden parts of the supports above the ground so they could drain and dry after rain, and thus be less liable to rot. The quarter bars transferred the load of the mill to the piers, and the post was held vertically – independently of the weight of the mill – for easier turning.

Better carpentry made it possible to increase the length of the main post and raise the height of the mill, to get stronger winds for the longer and broader sails made possible by the use of the whips, but it was difficult to stop the mill. It had to be pushed out of the wind to get the sails at right angles to it. In the early part of the fourteenth century, some clever carpenter or miller invented the brake blocks for the brake wheel, so stopping was much more controllable. Higher posts also lifted the whole mill off the ground so storage space became available underneath the body. To protect this area from the weather, round houses were built under the house or buck, and to help stabilise rotation, rims were added to the top of the wall and rollers to the bottom of the buck. The two forms of post mills are shown in Figure 5.

In the Brittany area of France in the late fourteenth century, a short tower mill began to be built. The lower storey was of stone and the second of wood, with a horizontal wind shaft in a conical cap that was turned by a tail pole. The stone part was waisted, and it was a very distinctive local type known as a *Moulin à Hourdais*, which does not seem to have spread outside the region. Nevertheless, it was the first tower mill, and a French manuscript of the fifteenth century shows a stone tower mill whose design again remained local. The Dutch began using windmills to drain the polders, and are credited with the first smock mill early in the sixteenth century. The first brick tower mill was built in the same century in the Flemish region of Europe. These designs did become accepted as the way forward and spread to England, causing the dominance of the post mill to decline in both Europe and England.

Mills were also built to drain the wetlands in England, and again it was the big landowners that financed the construction in return for the lion's share of the drained land. Disputes continued to be frequent over the milling toll, so in the reign of Henry I the toll grain had to be taken only by the King's Standard Measure. There was also the Assize of Bread to try to ensure that the miller acted honestly, and he could be pilloried for taking too much toll grain. He was also banned from keeping pigs or poultry so that he would not be tempted to feed them from his customers' meal.

Probably in the middle part of this period, the common sail evolved. Its angle to the wind was retained at 15 degrees, but all of the sail area was on the trailing side of the arm, with the sail cloth held by ropes on the front of the frame so there was only one sail per arm to reef, and it offered a smooth surface to the wind to improve power taken from the wind. Of necessity, mills had to grind over a range of wind speeds and hence a variable speed of rotation of the stones. This requires the feed of grain to vary with the stone speed, which would have been done by the eye and hand of a miller's helper. An early form of mechanical feed control was the 'clapper'. A piece of wood, fixed to the runner stone, banging the underside of a sloping trough from

POST MILL

POST MILL WITH
ROUND HOUSE

Figure 5. Post mill, post mill with roundhouse.

Figure 6. Smock mill.

the hopper causing the grain to jerk into the eye once in each revolution. Such feeding was uneven, so, at some stage, the damsel was invented. It is possible, and even probable, that this device was first used in watermills on under driven stones long before this time, but during this period its use became well established. The damsel was an iron rod fixed to the mace with three or four side bars, that shook the trough or shoe three or four times in each revolution of the stone. The feed hopper pivoted at the back to allow the shaking movement and a wooden spring held the shoe in contact with the damsel. This more rapid and gentler system gave a smoother and steadier feed of grain to the stones. The higher the speed of sails, shaft and stones, the faster is the hopper shaken, so that more grain is fed into the stones. The feed rate is therefore automatically controlled by the speed of the stone. On over driven stones a square section on the drive shaft served the same purpose and made the same noise, also sometimes called the damsel.

1301   Marton in Cleveland Mill rented for 13s 4d per year.

1306   At Hanworth in Norfolk, a new millstone was bought for £1 11s 5d.

1328   Shrewsbury mill soke was abolished. It needed an Act of Parliament.

1331   The Abbot of St Albans gained a legal victory to enforce his soke rights and confiscated almost a hundred hand querns from the locals. To add insult to injury, he then used them to pave a floor in the abbey.

1338   A tripod mill is shown in a painting now held by the Bodleian Library in Oxford.

1348   The population of England would be about 4 million when the first outbreak of the Black Death struck. Within a year or so, almost half the people had died. Villages and their mills were abandoned, and for the next 100 years mill building was a rare event. It would be 1600 before the population recovered to its pre-Black Death level.

1349   An engraving in St Margaret's Church, King's Lynn, shows a post mill.

1350   Mills were expensive to build and run, as the following example of rents shows:

Pasture	3*d* per acre per year
Farm	5*s* per year
Windmill	15*s* per year

1368   60lb (27kg) of iron, costing 5*s*, were needed for a new rynd and spindle.

1381   The Peasants' Revolt occurred against the imposition of the third Poll Tax to pay for the long and unending war with the French. Added to it also was the long resentment of the toll on milling and the Abbot of St Albans, who had had his levy legally enforced earlier, suffered an attack by the local people who dug up the abbey floor of their confiscated hand querns.

1414   The Dutch used a windmill for drainage for the first time.

1420   A French manuscript depicts a tower mill.

1430   In Coventry, the milling toll was fixed at a cash payment of 1*d* per strike, which was about a third of a bushel in an attempt to prevent the millers from cheating.

1557   Kings Mill, Liverpool. The soke was enforced by the Royal Command of Queen Elizabeth I.

1526   A Dutch wipmolen (hollow post mill) was replaced by the first smock mill (Figure 6).

1558   A sail was bought for 3*s*.

1580	Wind pumps were used in an attempt to drain the southern Fens and in Holland, long tail poles were first fixed to smock mills for turning the cap into the wind.
1586	Four or five ships of grain, destined for London from Essex, were pirated by Dunkerkers (the French).
1600	By this time, and possibly as early as 1550, a tower mill had been built in the Flemish region of Europe. This one would be copied and become the familiar type shown in Figure 7.
1604	The Dutch began building tower mills with eight floors, 100ft high and 30ft diameter at the base. A balcony at the forth floor was needed to operate the tail pole, and the lower two floors were the home of the miller.
1615	A post mill was sold for £47.
1617	The first patent was granted, establishing that ideas and new ways of doing things had a value. In later years, the addition of explanations and drawings would help in tracing and understanding the evolution of the windmill.
1621	The first post mill in America was built at Jamestown in Virginia.
1627	Pitstone Mill, Buckingham, was built.
1630	The Dunkerkers were still pirating grain ships from Norfolk, Suffolk and Essex. Charles I invites the Dutch Engineer, Cornelius Veymuyden, to devise a scheme to drain the Fens that would use Dutch expertise of wind pumps.
1635	Dee Mill, Chester. The Soke Law was upheld by the King.
1646	Dee Mill, Chester. The Soke Law was revoked by Parliament after the fall of Chester in the Civil War.
1665	Outwood Common Mill, Burstow, Surrey, rented for 5s per year.
1674	Patent No.174 was granted to John Johnson by Charles II. 'A new way … of mocon for a windmill … never yet practised … to the great advantage of the public, at a lesser charge and in a more convenient manner than any other yet practised.'

Alas, there is no more explanation or drawings, so it will never be known what this brilliant invention was, but it shows that changes were being thought about.

1675   Appledore Mill at Ashford, Kent, was rented for £6 per year.

1684   Patent No.243 was granted to Nathan Heckford, who was a draper of Essex. His was a horizontal windmill, but without drawings or technical text it is not possible to determine how it worked.

1700
to 1800   The post mill had passed the zenith of its development. Tower and smock mills with three to six pairs of stones were being built, but the challenge from steam was about to begin. Townships and independent millers were able to operate, and mill sokes were being challenged or ignored. Scientists began to quantify wind speed and engineers introduced cast iron into mill building. Patenting of new ideas introduced the fantail, for automatically putting the mill into the wind, and the horizontal mill, to deal with winds from any direction. Several new designs of sail to spill wind and the governor to control stone gap also appeared.

As the Industrial Revolution gathered pace, the strong new material called cast iron became available for gears, wheels, racks and the cross to carry not only the traditional four sails but also five, six and eight sails. By the end of this century, even though the first all-steam mill had been built, windmills (and watermills) were multiplying in number to remain the dominant producer of flour to meet the demand from a population increasing from 6 million in 1700, to 9 million in 1800.

England, with Ireland, could grow all its own grain, but after the middle of this century, increasing quantities of grain were imported from Prussia and France (when we were not at war with them). The windmills and watermills were still important at the end of this century, so much so that the Board of Agriculture had the famous engineer, Thomas Telford, report on the state of the art of these two sources of power.

Despite the wars with France, the fame of the French burr stone for milling fine flour reached our shores by the 1770–1780s, and a few began to be imported. However, since England fought the French almost continuously

Figure 7. Tower mill.

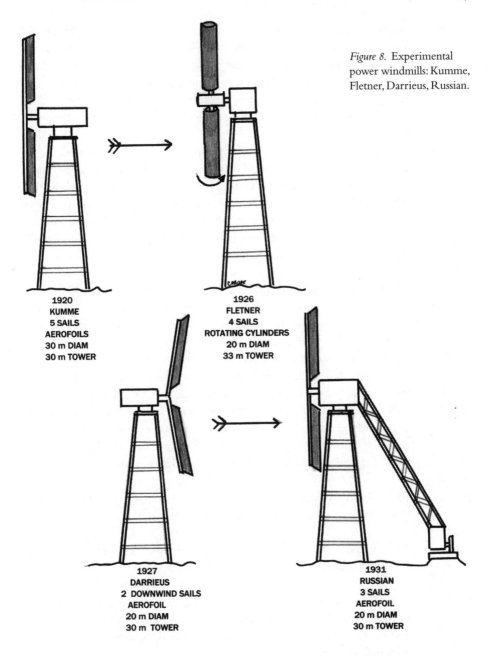

Figure 8. Experimental power windmills: Kumme, Fletner, Darrieus, Russian.

1920
KUMME
5 SAILS
AEROFOILS
30 m DIAM
30 m TOWER

1926
FLETNER
4 SAILS
ROTATING CYLINDERS
20 m DIAM
33 m TOWER

1927
DARRIEUS
2 DOWNWIND SAILS
AEROFOIL
20 m DIAM
30 m TOWER

1931
RUSSIAN
3 SAILS
AEROFOIL
20 m DIAM
30 m TOWER

from 1793 to 1815, few of the stones or the quartzite to make them would have got here in that period. England, therefore, continued milling mainly with the local grit stones.

1703   There was a great storm and 400 windmills were wrecked, mainly by being tail winded, causing some or all of the sails to be blown off. In some cases, the wind shaft may have been blown out and post mills toppled. Where the sails kept turning despite the brake, the friction between the wheel and the shoes would generate heat and sparks that could set the mill on fire! There would follow a big rebuilding programme, but such a programme could only use existing technology. The disasters must, however, have spurred on the builders to find new ways of keeping into the wind and reefing them.

1708–9   George Sorrocold built a water pump at Clerkenwell in London. A rudimentary sketch by Sutton Nichols in 1712 shows it to be a tower mill with six sails that were two to three times wider at the tip than the heel. It was an early tower mill, and probably the first with six sails that must have been mortised through the wind shaft. A gale in 1720 blew off the sails and cap, which probably deterred others from building more with six sails until the advent of the cast-iron cross some fifty years later.

1712   Thomas Newcomen invented the beam steam engine and sold the first to pump water out of a coal mine near Dudley Castle in the Midlands. It could not drive rotating machinery and wind millers would not have heard of it – but their great-grandchildren certainly would!

Windmills were still all-wood constructions with only a few iron pins and straps. They were turned into the wind by poles on the buck or cap by the miller's brawn. The use of a wooden rack on the curb of a tower mill, to engage a wooden cog that was turned by a loop of rope reaching to the ground from a large-diameter counter wheel, was introduced about this time. The small cog on the long rack meant that the effort needed to move the heavy cap and sails was much less than pushing on a pole.

1720   Jethro Tull invented the seed drill for more even and rapid planting to improve the yield of grain.

1726   The earliest smock mill that was dated was erected at Wrattling, Cambridgeshire.

1736   James Watt was born in Greenock in Scotland, and as an adult he became known as the father of the steam engine; it would be these engines that would, a century or so later, bring about the decline and demise of the windmill for commercial uses.

1738   George II grants patent No. 561 to John Kay of Lancashire for an 'upright' windmill. From the text, the vertical sails open and close like doors so that it will operate from any wind direction without needing the miller to push the mill around. It probably did not work, but the problem of keeping the sails into the wind was being seriously studied.

1744   Another patented attempt by William Perkins at new windmill technology. This system had eight horizontal sails, opening and closing by weights during rotation so it would work from any wind direction. The sails were clothed at both sides like an aerofoil and rotation could be reversed.

1745   Turning the sails into the wind was still done with poles, levers or hand-turned gears, but in this year Lee patented the fantail – a five, six or eight-vaned wheel fitted to the tail pole at ground level and at right angles to the main sails, so that when the wind veers, it strikes the fantail vanes. The vanes rotate and drive a spiked wheel on a plank track on the ground and turn the cap and the sails back into the wind. Being behind the mill, it would take a large shift in wind direction before the vanes operated, but it was designed to be an add-on to the normal windmill and was therefore more understandable, acceptable and feasible to the milling industry.

1749   A patent was obtained by Richard Langworthy based on the original vertical wind mills of the Afghan region. To make it work from any wind direction, a three-quarter cylinder was turned round the sails and into the wind by a large fin so that the quarter opening always faced the wind.

1750   Andrew Meikle invented the fantail fixed to the cap itself with a worm gear engaging a toothed rack on the curb to turn the sails back into the wind. This fantail was on the horizontal extensions of the main cap support, giving a better response to changes in wind direction, but still in the lee of the cap. He did not apply for a patent.

1751   The introduction of cast-iron gears, by John Smeaton, and the invention, by Carnus, of bevel gears and curved faces for cogs to ensure constant meshing, considerably improved life over wooden gears to improve the smooth running of any machine, including windmills.

1755   John Smeaton designs a seed crushing windmill for oil production in Wakefield. It had a cast-iron rack on the curb and a cast-iron pinion, operated by a chain wheel, to turn the sails into the wind. It also had a cast-iron wind shaft and cross for four sails.

1759   John Smeaton was, by now, a very eminent engineer, who had built Eddystone lighthouse and had become a Fellow of the Royal Society. He carried out a very detailed study of the two sources of power available at that time by constructing models of water wheels and windmill sails. The results were read to the Society and published in Philosophical Transactions 1759–60, vol.51, pp 100–174. To study the behaviour of windmill sails he built a machine using sails 450mm long. Like a modern wind tunnel, the machine could provide, albeit for a short time only, a steady wind whose speed could be calculated and selected. The shape, size and angle of the sails could also be changed. The performance of a sail was measured by multiplying the number of turns the sail made by the weight it lifted on a cord wound on its wind shaft. Using this 'product', the behaviour of different sails at different wind speeds could be compared. It was a highly scientific, logical and objective investigation, and his main conclusions were:

Maxim 1   The speed of rotation of the sail is directly proportional to wind speed.

Maxim 2   The maximum load on the sail is almost proportional to the square of the wind speed.

Maxim 3   Combining 1 and 2, the power from the sails is almost proportional to the cube of the wind speed.

Corol 1   The power from the sail is proportional to its area.

His experiments were with four sails at a theoretical setting, and at standard Dutch and English designs, and some tests with eight sails but no tests with five sails. The comprehensive document of results was available to engineers in industry, to whom he would be well known and respected, but he would have been a stranger to the builders and operators of corn mills, who would have been very reluctant to accept ideas based on a toy played with by a man from outside their trade.

1761   The Bozeat post mill, Northamptonshire, was built. It was unique in design, as the brake wheel was behind the wallower which drove two sets of stones via a spur wheel as in tower mills. It could grind 20 quarters of grain per day (4.5 tonnes).

1768   Andrew Meikle and Robert Mackell got patent No.896 for a rotating grain cleaner with fans that removed small stones, dust, mould, chaff, seeds and very small grain, so that clean reasonably uniform-sized grain could be fed into the stones of the mill.

1770   John Bond invented the iron Suffolk plough, to get faster preparation of the fields for sowing wheat.

1772   Andrew Meikle invented the spring sail. Shutters of wood or canvas on a frame were pivoted by an operating lever that linked all the shutters on a sail by a wooden slat or an iron rod. A spring pulled on the rod, so closing the shutters. Sudden gusts pushed the shutters open, spilling wind, thus avoiding an over speed incident. When the gust had passed, the power of the spring closed the shutters for continuing steady operation. The spring tension could be adjusted to change the force at which the shutters opened, which is the equivalent of reefing. The springs on each sail still had to be individually set, requiring the mill to be stopped four times just as for common sails. Again he did not patent the idea, which would become a very successful way of dealing with changing winds and gusts.

The Jeffrey map of Yorkshire shows over 100 mills.

1774   Perhaps no one had taken much notice of John Smeaton's work! In this year he decided to build his own flint crushing mill, and it was not conventional. Further work on his model must have lead him to believe that four sails was not the best, as his mill in Leeds had five, and only a cast-iron cross made this configuration possible. It is also probable that the sails had triangular leading edges and were twisted to the optimum, as determined in his tests. Windmills with more than four sails never did become widely accepted, but the cast-iron cross would be widely adopted whatever the number of sails. An alternative for four sails was the canister, two square cast-iron tubes at right angles, into which were wedged the sail stocks, and the canister itself was wedged and bolted to the end of the wind shaft. Both were seen as a better practical alternative to the mortising of the sail stock through two square holes in the poll end of the wind shaft, which had become a critical point of weakness and was limiting the length and breadth of sails. The cast-iron cross was Smeaton's real legacy to the windmills.

1780   The tower mill at Etton in the East Riding of Yorkshire was fitted with a rack and pinion, to operate the movement of the cap by a wheel and chain to the ground instead of the double tail pole.

1783   Benjamin Wiseman takes out a patent for a vertical windmill with sails like the jib sails of a ship. It could work in any wind direction and even had a centrifugal release system to free the sails in the event of a strong gust.

1785   Robert Hilton gets a patent for roller reefing sails that would replace common or spring sails on the normal windmill. It had a very complex operating system that was unlikely to work.

1787 Benjamin Heam patents a sail system for dealing with variable and gusting winds. The sails pivot on the sail arms and are held into the wind by ropes fixed to a rod that passes down the complete length of the wind shaft. The pull on the rod is maintained by weights that can be changed while the mill is operating. It looks like a good, simple and reliable system that, alas, was never built.

1788 The imaginative Andrew Meikle had another good idea, and this time obtained patent No.1,645 for the first threshing machine. The wheat was fed in ear first, and beaters moving at 2,500ft/min separated the ears from the straw, which was removed by revolving rakes. Fans blew out chaff and dust to provide a cleaned grain for the mill.

1785 Albion Mills at Blackfriars in London was the first designed and built to use steam power for milling. Phase 1 used a single 50hp Boulton & Watt condensing engine with another of Watt's inventions – sun and planet gears – to get rotary motion for ten pairs of stones. (The biggest windmill in England at the time could only power six pairs of stones.) When a second engine was added in 1788, there were twenty pairs of stones in production and there were plans to add a third, raising capacity to thirty pairs. Alas, it was completely burned down in 1791, so the miller had nothing to fear from these new fire breathing mills! Or had he?

1787 Thomas Mead patents the ball governor for automatic feedback control of the stone gap as the wind speed changes. Andrew Moulton, an associate of James Watt, saw this device in use on the stones at the Albion Mills and, soon after, Watt produced a similar device to control the flow of steam to his engine to maintain constant speed and make the steam engine a more attractive and controllable source of power. It is interesting to note that, while Watt took out many patents and is credited with inventing the first speed feedback control, he did not take out a patent on this device, possibly acknowledging the prior invention of Mead.

1788 A little before this date, at Margate in Kent, Captain Stephen Hooper had built a horizontal mill which had a wheel of 12m diameter with 8.5m-high boards that were flat to the wind on the drive side but rotated to be parallel to the wind on the return side to reduce drag. As on any horizontal mill, only half the sails provide power at a given time – hence the large size of the wheel. In this year, another Hooper horizontal mill was built at Battersea in London, but the maintenance of the board rotate and pivot system would cause serious problems. Two more were built at Hounslow and Mitcham Common in Surrey, but all had ceased operation by the early 1800s.

1789   Captain Hooper, however, did not restrict his interest to horizontal mills, and patented his version of roller reefing where the rollers were at right angles to the sail whips, reefing short sails operated by a rod that passed down a central hole in the wind shaft. Movement of this rod by hand could furl and unfurl the sails while they were turning, but the system could not spill wind in gusty conditions. It was used on several mills in Yorkshire, but again maintenance was a problem, particularly with the ropes fixed to the end of the sails that did the reefing.

1792   George Silvester is granted a patent for a vertical sailed windmill where the sails are carried by, and will pivot in, a square section chain. The wind on the sails moves the chains which turn two wheels which can drive machinery. The chain/sail assembly is kept into the wind by a large vane, and the eight sails open and close like doors as they move round the wheels. It was cumbersome and complex, but it is possible that the Duke of Bedford, who promoted technical innovation on his estate, had one installed at his Crawley clay kiln site to drive the 'pug' mill. Fortunately for us, the artist, Thomas Fisher, painted the scene about 1820, showing a horizontal windmill. It was expensive to maintain and was eventually removed about 1830. None were ever used for corn milling.

1796   At Chislehurst in Kent, the parish mill cost £1,765 to build. This was a very large investment for the time. In Hull, a windmill for paper production was built, with five roller sails with air breaks on the leading edge. The Suffolk drill was invented and it could sow fifteen rows of seed evenly to further increase the yield of corn.

1796–8   The continuing wars with France made grain and flour imports almost impossible, so the home production of flour was vital, and such production was reliant on windmills and watermills. Sir John Sinclair, President of the Board of Agriculture, commissioned the young engineer, Thomas Telford, to do an in-depth study of both. Telford went to Shrewsbury to work with William Hazledine, who had been building mills for some ten years. The documents and drawings that Telford produced are now held by The Institute of Civil Engineers and an edited version of them was published in the *Transactions of the Newcomen Society*, 1936–37, vol. XVII, pp 205–214. With all the innovations on sail design and luffing going on, these remarkable documents show what the experienced millwright was actually installing at this time. The data is so comprehensive that it would be possible to build a windmill from them. The main characteristics of such a mill are given on the following page:

HAZLEDINES TOWER MILL

Height	11m
Floors	4
Curb diameter	5.3m
Base diameter	7.6m
Cap Type	Ridge
Luffing	Counter wheel turned by rope to ground
	Cast-iron pinion engages cast-iron rack
	Gear ratio 144:1
Wind shaft	Wood. Inclined at 8 degrees
	Sail stock fixing. Cast-iron canister
Sails	
Rotation	With the sun, i.e. clockwise
Type	Common, with wind board
No.	4
Arm length	10.5m
Sail length	9.1m
Sail geometry	Tapered, 2.6m at heel, 2.1m at tip
Swept area	343 sq m
Total sail area	85 sq m, 25 per cent of swept area
Weather angle(twist)	18 degrees at the heel
	5 degrees at the tip
Wheels	
Brake	2.8m diameter, compass arm, wood, 62 teeth
Wallower	1.6m diameter, compass arm, wood, 44 teeth
Great spur	2.0m diameter, compass arm, wood, 68 teeth
Stone nuts	0.5m diameter, solid, wood, 17 and 18 teeth
Sail to stone gearing	6.18 or 5.83
Brake	Wooden shoes
Tentering	Manual screw, no governor
Stones	
Type	French burr
No.	2
Diameter	1.36m and 1.46m
Rotation	Against the sun, i.e. anticlockwise.
Speed	93 and 87rpm at sail speed of 15rpm.
	Rim speed, 14.7mph for both.
Dressing of the stones	
Furrows	Straight
Harps	10
Furrows per harp	4
Furrow geometry	Tangential to left

The windmill, complete and ready to operate cost, £500. About £250,000 in today's money. The fantail, patented in 1745, spring sails, invented in 1772, and the governor, patented in 1787, were not used, showing how long the new ideas took to be accepted. Local grit stones were available at about £4 each at the quarry, to be used as the bed stone with a burr runner. Peak stones from Derbyshire are not mentioned, so it may have been that the transport costs from Hathersage to Shrewsbury, some eighty miles, did not justify their use. However, importing burr stone lumps from France (despite the war) and making up the mill stone from them obviously was cost effective, even though such stones cost £26 and £30 for the 1.36 and 1.46m-diameter stones.

1797   Steam power was having almost no impact on the wind millers' flour trade and even Richard Trevithick, one of the Cornish pioneers of steam power for mine pumping, built a wind-powered pump to drain the Ding Dong Mine at Madron near Penzance.

1800s   Flour milling, feed milling and drainage continued to be the main – and growing – uses of mills; mineral grinding, seed crushing and sawmills added a few hundred more. The mills, mainly brick tower, were getting larger and taller, with more floors for storage and flour treatment machinery and more height to use the stronger winds. By now, twist in the sails was an accepted development, as was Smeaton's iron cross or the cast-iron canister, to make strong joints for longer and broader sails. The problem of dealing with gusty winds and on-wind sail adjustment was solved at the beginning of the century, when wind and water dominated as power for industry and steam was still in its infancy. The miller got his payment for milling in coin or flour, which at this time was 1s 6d per comb and a further shilling for dressing. With mills making from 500kg to 7.5 tonnes of flour per week, the mill owners were wealthy men, making £100 to £750 a year, in the same range as small land owners and senior clergy.

This is much more than the £20 to £25 a year earned by farm labourers who, during the French Wars, were paying 6d for a 4lb loaf. The wars ended in 1815, but the Corn Laws protected the farmers from imports of foreign wheat. At the same time, other trade was resumed with France, and the import of French quartzite for burr stones began and were soon the dominant material for milling wheat.

Wheat yields increased from around 30 bushels per acre in the early years of this century, to about 55 bushels by the 1860s, but with the population growing faster than in the previous century, import of grain began to increase once the Corn Laws had been repealed in the 1840s.

The multi-bladed windmill appeared in England, but was exploited principally in the Wild West of America. Grain arrived both from our Empire and America to be ground in port-based steam-driven mills using the new roller process developed in Central Europe in the middle of the century. In the 1880s, there were 10,000 windmills in England, but by the end of the century these changes had halved the number of windmills.

In Holland, by 1850 there were 9,000 windmills milling corn or draining polders.

Ironically, it was the introduction of large steam drainage pumps from England at this time that began the decline of their windmills and reduced them to about 4,000 within fifty years.

Another newfangled energy also appeared – electricity – and a model windmill turned a generator and lit a few lamps!

1801   Wrawby post mill was valued at an annual rent of £13 9s.

1802   Grinding and dressing of 1 bushel of wheat cost 9d.

A new mill at Foulness in Kent cost £536.

At Sutton St Edmunds, a 12hp steam-driven pump was installed to drain 4,000 acres, and in the Isle of Ely, 230 wind pumps began to be replaced by steam pumps. When complete, only nineteen steam pumps were needed.

1804   Bywater obtained patent No.2,782 for roller reefing of cloth sails. The full power of the wind rotating the sails was used to roll up or out the longitudinal cloth across the width of the sail frame, and it could be done on wind by control lines that passed down the wind shaft and the floors of the mill to the miller. It was not automatic, so the miller had to respond to gusts. A partnership was set up to market the system, but it was dissolved in 1807 with only a few sold.

1806   Hessel Mill near Hull was built, with five roller sails and air brakes to grind limestone for whitewash. It is worth noting that these unconventional designs were still not on corn mills. It also had another innovation and that was a fantail on an inclined frame so that the fantail was above the level of the cap and able to respond quickly to small changes in wind direction (Figure 7).

1807   William Cubitt patented his method of sail control 'on wind and rotating', i.e. automatic adjustment for varying and gusting winds. In patent No.3,041, published in June, Cubitt had ingeniously combined the shutters of Meikle with through shaft operation and added the critical ingredients of:

1   A mechanically reliable linkage from shutter to shaft.

2   A weight system held the shutters closed for normal operation, but they opened automatically to spill wind in gusts. If the wind strengthened, partial opening helped to give a more constant power to the stones and keep them turning at a more or less steady speed for milling. The weights could be changed at ground level while running so control, start up and shut down were better, safer and quicker. The patent was granted for fourteen years.

In the same year that the patent was granted, Cooke's Mill at Stalham was built with Cubitt's sails for £800, and its contents insured for £200. Horning post mill also changed to Cubitt's sails. Both had sails 8.5m long and the fee for use was initially determined on this parameter, but was soon rationalised to £25 per mill. (Cubitt married Cooke's niece in 1809.)

1809   The tower mill at Oundle (built in 1739) with two pairs of stones was sold for £740.

1812   Southdown Mill was built at Great Yarmouth. It cost £10,000 and was 14m diameter at the base rising to 30.5m at the curb, with sails 26m diameter peaking at 46m. The wind shaft alone weighed 5 tonnes and the sails and cap 15 tonnes. It had six pairs of stones and could mill 200 quarters (50 tonnes) of flour in a week. This was the answer to steam! Remarkedly, it was not the highest. That distinction goes to Bixley Mill, Norwich, the tower of which was 42m high.

However, further south, at Sutton St Edmunds in the Fens, steam was first used for drainage. A 12hp beam engine was installed to drain 4,000 acres, and in the Isle of Ely the replacement of 230 wind pumps began. When complete, only nineteen steam engines were needed. Steam was fighting back.

1813   About twenty mills were now using Cubitt's sails and Cranbrook smock mill, the tallest mill in Kent – height 22m with seven floors – was built, with Cubitt's sails and a cap fantail for £3,500.

1814   A smock mill was built at Boxhall, Suffolk, for £528, and, very significantly, another all-steam mill was built at Lincoln Street, Hull.

1815   There had been a long war with France which had just ended, but to protect British grain growers the Corn Laws were introduced to stop imports, so the home mills exclusively milled local grain.

1820   At Ten Mile Bank in the Fens, seventy windmills were made redundant by one steam pump, and a further seventy-five wind pumps ceased operation at Littleport Fen when a single 28hp steam pump came on line. Steam was answering back again.

At Buzza Hill in the Scilly Isles, a Mediterranean-type mill with two storeys and conical cap on the parallel-sided stone block tower was built. It had eight jib-like sails with rope braces to an extended wind shaft.

1821   Cubitt's patent expires, but the performance of the system ensures its continued wide adoption both for new mills and replacement sails on existing mills. The sails became known simply as 'patent sails'.

The advantage of the better response of the inclined fantail was also being recognised, and it was being used for new mills. Act 36 of George III legalised payment of the knaveship or levy for the grinding of corn in coin, but did not specify the amount, which was still left to local tradition.

1834   In Hungary, Jakob Sulzberger, unheard of in England, invented the roller milling process for flour milling – but what had this to do with windmills?

1838   Henry Chopping, son of the miller at Matching, Essex, set up a little windmill with many small sails on an annular ring. A radical departure from anything at the time, which he was proud to show to visitors to his father's mill and was the first record of a multi-bladed windmill.

1840   About this time, in America, the farm windmill began to appear. It had small fantail-like vanes on an open wooden tower to pump water for the pioneers as they went west to settle in the plains.

1843   The last mill soke in force in Jedborough was legally abolished.

1846   The Corn Laws, introduced in 1815 to protect the British wheat industry and prohibit corn imports, were abolished. Foreign wheat could be imported and would arrive at the ports, which were a long way from most windmills, but could the new varieties of corn be ground on old-fashioned stones? Only time would tell!

1848   Cooke's Mill at Stalham, the first to be fitted with patent sails, was converted to steam power. Coal was cheap, at less than £1 a tonne, and with it the miller could mill when he wanted to, and not just when the wind blew. A foretaste of things to come!

In Switzerland and Hungary, fourteen years since its invention, full-scale evaluation of milling with steel rollers began. Wind millers would be blissfully unaware of this ominous event, or its pending disastrous effect on their trade.

In Holland, three 400hp steam pumps from England were delivered to drain Haarlemmermere. Their success began the decline of wind pumps and windmills.

1849   A court case in Wakefield in Yorkshire fixed the toll at 1 to 16, which could be used as a legal president through out the country.

Swinegate Mill, near Dover, needed 40,000 bricks for its tower.

1850   H.G. Magnus, a German physicist, discovered that when a wind blows on a rotating cylinder, a sideways thrust is produced on the cylinder.

In Holland, windmills and pumps reached their peak of about 9,000.

In the four years since the repeal of the Corn Laws, 25 per cent of grain was imported.

A pair of monolithic Peak stones cost £5, while a pair of burr stones would be £36.

1854   On 28 March of this year, at the cottage of the Holderness Road Mill in Hull, the first son was born to the proud miller. The baby's name was Joseph Rank. Over the Atlantic in America, David Halladay invented the multi-sailed farm windmill with self-reefing in strong winds or gusts. The fame of Henry Chopping's toy would not have got to America?

1855   At Haverhill in Suffolk, a Mr Ruffle built a 20m high tower mill with an annular ring of small blades based on Henry Chopping's original model. It had a 15m-diameter wheel with 120 blades, 1.5m long tapering from 350mm to 300mm. There would eventually be four of these mills in Suffolk.

1860   Moses G. Farmer in America patents windmills with dynamos to power a few primitive light bulbs and even makes a small model to prove it. Another laughable idea from an eccentric scientist! Or was it?

1867   In America, the Revd Wheeler (a very suitable name) invented the side vane for farm windmills that automatically turned the sail wheel out of the wind to control gusts and stronger winds. This became as popular as the Halladay method for farm windmills that were being installed in their thousands.

1868   Henry Chopping (and Warner) take out a patent on a mobile windmill with automatic reefing of multi-bladed annular sails.

1870   Full-scale commercial roller milling with inter-stage sieving was well established in Austria, Hungary and the USA on their hard grain, which stones had difficulties in milling. It also produced a whiter flour at competitive prices which began to reach the British market, but because of the continued increase in population, the demand for flour was also rising, which meant that windmills were still being built.

1871 A new cast-iron brake wheel for St Peters Mill, Broadstairs, Kent, cost £62.

1873 Charles Edwin Hammond patented a device to control the setting of patent sails as well as the weights. A governor, driven from the wind shaft, caused the striking rod to move in or out and control the position of the shutters to keep the sails turning at a preset speed in varying winds, so the stone speed, and hence the flour production, were done at steady conditions. In 1876, he bought the Jack and Jill windmills at Clayton, East Sussex, and installed the system on the sails of the tower mill, Jack and also on the patent sails of the nearby Herstmonceaux post mill. A good idea, but with mills facing ever increasing competition from imported grain and roller milling, no other mills were equipped.

1874 Joseph Rank buys his first windmill and struggles to make a living. It is said that he tried soaking hard imported grain to soften it for stone milling – but without success.

1875 –1879 Very poor harvests resulted from wet cold summers. Imports of grain rose even more to meet the deficit, and more port-based roller mills were installed.

1879 Rank rents a gas engine-driven stone mill on Wednesday, Thursday and Fridays only. He gets his orders in the early part of the week, and with reliable output at lower cost, begins to take the local Hull market. He soon adds two roller mills, generating more sales, profit and capital for further expansion. Even so, the stalwarts of the windmill still had faith and built Shipley smock mill for £2,500.

1880 A survey of the mills in East Yorkshire gives a total of seventy-nine mills. Sixty-two had four sails, of which fifty-six were patent, five had roller sails and one common. Fourteen had five sails – two with roller and twelve with patent.

1883 In America, Thomas Perry did a very exhaustive study on the performance of farm windmills that was as detailed as Smeaton's work on the English windmills in the previous century. The results of Perry's work was the much more efficient curved metal blade that could be mass produced by stamping out of sheet metal.

In England, Henry Chopping puts a farm-mill-type sail on his mill at Roxwell after the conventional four sails were blown off in a gale.

1885 Rank builds Alexandre Mill at the dockside in Hull. It uses only rollers, works two shifts and uses a blend of local and imported grain from USA and India. It had a capacity of 1.6 tons of flour per hour and could compete on price and quality with flour imported directly from the States. The local wind millers were not so fortunate!

The farm windmill was being built in England for water pumping and for export to the colonies, not only for pumping but also for corn milling.

1887 The Association of British and Irish Millers complains to the British Government about the dumping of American flour on the British market with the consequential closure of not only windmills, but also some new roller mills as well. Its report states that the population increased from 18 million in 1851 to 26 million in 1881, but workers in mills fell from 36,000 to 23,000 in the same period. In 1879, 10,000 windmills were working but now (1887) there were only 8,814. An additional 460 roller mills have been opened, but flour imports increased by 300,000 tons in 1886 alone so that 75 per cent of all bread was now made from imported grain. The Association concluded that the whole flour milling industry was doomed. Alas for the windmill they were right, but thanks to the efforts of Joseph Rank in particular, home production of flour was assured.

In Scotland, Professor James Blyth built a horizontal windmill to generate power for remote places. It was rated at 3kW DC, but failed to be commercially exploitable.

1891 In Denmark, Professor Poul la Cours fitted DC generators into conventional windmills for village use. At about the same time, in America, Charles Bush added generators to the farm windmill for 12 volt DC battery charging for remote single homestead use. Both systems could only power a few lights, but wind would never be used for large-scale generation of power – or would it?

1892 Two new sails cost £42 for the Adisham Mill at Canterbury.

At Much Hadham in Hertfordshire, Richard Hunt, who had been a miller and successful merchant, retired to this village at the age of seventy-eight and decided to build another mill. It had eight storeys, eight double-sided patent sails and was 35m at the curb, only a few metres less than the mill built at Southdown in Yarmouth, and only the seventh mill to have eight sails. All the machinery was new, including the four pairs of burr stones, and the cost was £5,000 – about £350,000 in today's money. With established mills closing in their hundreds, it was a foolhardy undertaking that would be unable to compete with larger roller mills, and it was the last mill ever built in England. He also had to contend with the variability of his source of power – the wind. A gale in 1895 blew off three of the sails and the mill never worked again, so the mill was not only the last, it also had the unenviable distinction of operating for the shortest time.

1900 to 2000   The decline, closure and neglect of the traditional windmill had continued in the closing decade of the last century and the opening decade of this. The smaller wooden post and smock mills were first to go, while the brick tower mills struggled on with wind supplemented by other power. Some disconnected but retained their sails and milled only by steam, gas or oil engine power. In America, the farm windmill continued to be adopted on a large scale – still mainly for water pumping. They were now being built on high steel towers with sixteen to 100 blades of 3.5 to 4.8m diameter, capable of pumping thirty to forty gallons of water per minute, to a head of 7.5m for the homestead, its animals and irrigation. Many were also being installed in England, the largest of which had a 12m-diameter wind wheel developing 12hp in a 25mph wind.

Steam from king coal was still about £1 per tonne and accounted for some 95 per cent of power use in the opening decade, but its reign would decline with the coming of cheap oil and gas. The flying machine took to the skies and the First World War increased the pace of its development. The design of its propeller became the inspiration for new small windmills generating power, but the traditional English windmill had virtually ceased any commercial production by the end of the Second World War, and only a few enthusiasts were concerned about their disappearance and were trying to get some restored. The spread of rural electrification in the States had significantly reduced the numbers of many sailed farm windmills there.

The use of all types of windmills was dying! But the world energy crisis of the 1970s meant that renewable sources of energy were needed – and the wind is a very powerful and inexhaustible source of energy. The few kilowatt generators of the 1930s became the megawatt wind turbines of the 1990s.

1904   Southdown Mill, Great Yarmouth, that cost £10,000 when built, was sold for £100 and demolished.

1905   By now, 300 million bushels of wheat were being imported – all for the roller mills. Sixty million bushels of wheat were still produced at home, but most of this went to roller mills, leaving only a small proportion for the small windmills.

1910   Poul la Cour's windmill generators in Denmark were supplying 100 to 200 villages with power for lights. The towers had been increased to 24m in height with four 20m-diameter sails generating 5 to 25 kW, but they could never generate much more than this … or could they?

The last wind drainage mill was built at St Olaves.

1914   The First World War and the Government ruled that flour from windmills was unfit for human consumption due to the very poor standards of cleanliness that could be maintained in them, leaving the few remaining windmills only animal feed to mill, so more closed. Aeroplanes, invented by the Wright brothers only a decade earlier, were flying through the skies above Europe. Both sides wanted faster planes, and that required an understanding of aerodynamics to improve lift from the wings and higher speeds from the windmill at the front that propelled them. After the war, the aerodynamicists put their thinking caps on to see what could be done to the ground-based windmill, and in the next decade they came up with some weird and wonderful ideas.

1920   Road transport was now well developed and local roads were improving, and together these put good cheap flour into all parts of the country and more mills closed. A few carried on with sails stopped or removed, using small oil engines

The wind would produce 3 to 5hp to drive one pair of stones to grind 50 to 150kg of grain per hour. An oil engine of 10hp cost only £90 and would run for 12 hours a day driving three pairs of stones. They could, however, only grind local grain, so closures continued, so much so that the English windmills were in their death throes! Of the 10,000 in 1879, there were less than 350 left. But wind power had not died.

America had 2 to 3 million farm windmills and more were being installed. Their mechanical and electrically minded owners were adding second-hand dynamos from cars and lorries to charge batteries for just a few lights on the farm.

Kumme in Germany built a five-sailed windmill with 20m-diameter sails with a gearbox in the cap driving a central shaft to an electric generator on the ground (Figure 8).

1921   A specific example of the decline of the windmill is shown for East Yorkshire:

Year	Mills
1850	113
1892	104

1897	57
1909	33
1921	18

Only Skidby Mill near Hull now remains in sail.

1925   Fletner built a ship called the *Baden Baden* using the Magnus principal, discovered in 1850, whereby a rotating cylinder in a wind develops thrust, and sailed across the Atlantic.

About this time, in Holland, their millwrights were trying to make existing sails better. Using wing aerodynamics, Dekker added a sheet metal nose to the entire leading edge of the sail stock to reduce the drag caused by the 300mm-wide flat face. Based on maritime sails, Fauel designed a wind board for the leading edge that had a gap between it and the sail stock. The idea was that the partial vacuum formed in the gap would pull the sail forward, particularly in low wind speeds. Both types, known as 'improved sails', were so successful that air brakes had to be added to control sail speeds in high winds. Compared with common sails, which could drive the windmill or pumps 25 to 30 per cent of the time, the improved sailed windmills could operate for 50 per cent. Unfortunately, steam, gas, oil or electric engines have an availability of 80 to 90 per cent.

1926   Having sailed his ship, Fletner then built a windmill with four slim rotating cylinders as sails. Sail diameter was 20m on a 33m tower, and it could generate 30kW in a 25mph wind. It must have been a marvellous sight! (Figure 8.)

The Miles Mill at Faversham installed a 15hp Blackstone oil engine to drive three pairs of 4ft 6in-diameter stones.

1927   Marcellus Jacobs of Montana, USA, had been trying to improve the electricity generation for his farm windmill, which, like all the others, was very low powered. Possibly inspired by the wooden propellers on the early aeroplanes, he designed a 3 slender sailed wheel of 4.5m diameter on a 15m tower to drive a low speed 110 volt DC generator to charge a 21kW battery pack (Figure 9). This was the first major redesign of windmill sails since Cubitt's patent sail in 1807, and the fore-runner of the modern wind turbines of today, but at this point in time the generators were still small DC units that could not power the growing range of equipment made for 250 volts AC from the grid.

1928   Prof Albert Betz of Germany calculated that the maximum energy that could be got from the wind was 16/27, or 59 per cent, when the exit wind velocity from the sails had been reduced by one third of the entry speed.

1929   The French built a two-sailed windmill with 20m-diameter sails using a Darrieus design, with the sails down wind of the tower – the complete opposite of traditional windmills (Figure 8).

The English answered back with the last old-fashioned smock mill to be built. Holman Bros of Canterbury installed the mill for Sir William Beardswell to generate electricity at his St Margarets Bay house, near Dover.

1930   The American farm windmill reached its zenith of about 6 million units. Back in England, windmills were becoming derelict, and fewer than 100 were still in wind.

1931   Across the Channel, in Germany, there was a proposal by Honnek for a 50MW windmill. It was to be 360m high and have five sets of sails of 70m diameter (Figure 10). The estimated cost was £150,000 and the cost of power from it would be one farthing per kWh (this is equivalent to 0.001p – compare this with the price on your next electricity bill). They were able to build the Zeppelin airship, but not this windmill, so could power ever be generated on this scale by windmills?

Savonious developed a vertical rotor which was two half cylinders offset on the axis so the wind followed an 'S' path through them. For 30kW it would have needed a 100m-high tower with a 30m-diameter rotor (Figure 11).

The Russians built a 100kW three-sailed mill at Yalta on the Black Sea coast. It was 30m high and resembled a very large post mill (Figure 8).

1933   A single large revolving cylinder was built at Burlington, New Jersey, to quantify the magnitude of the Magnus effect and the costs of installing large units (Figure 11). It was abandoned, as were all the other large power projects conceived in the 1920s. Power could never be generated from the wind!

1934   Palmer Putnam, a wealthy American, wanted power for his large new country house with all its 'mod cons' that needed 250 volts AC. Grid power had not reached his area, and for one user it would be very expensive to run it to his site so, despite the very recent failures, he began to look at the windmill to provide his power. What began as a quick 'look-see' at what he could get, became a major in-depth study of the state of the art of the technology and the earlier failures of windmills for power generation, and how they could economically integrate into a power supply scheme. It took him three years to design HIS 1.25MW, two-sailed windmill for AC power, and two years to find someone to build it.

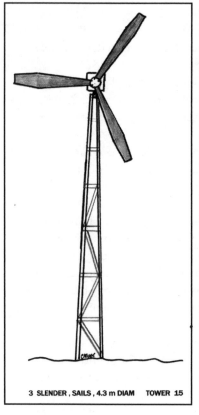

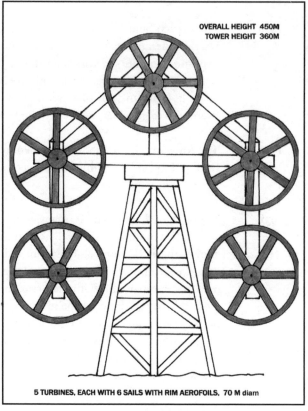

Figure 9. Jacobs Farm Windmill.   *Figure 10.* Experimental power windmills: Honnek.

1939   Palmer had got financial backers and an engineering team to check his concept, and the Smith–Putnam wind turbine was to be built to his original design (Figure 11).

1940   Wartime again, and the American farm windmill for pumping was heading for the happy hunting grounds as rural electrification swept across the mid-west, and where the wires had still not reached, a high voltage version of the Jacobs generator kept the flag flying for the farm windmill. The Smith–Putnam windmill for AC generation was being installed for evaluation. It had two slender sails, 53m in diameter, on a 36m steel tower with a 1.25MW alternator, and was built at Grandpa's Knob in Vermont. At home, no one had time for the ancient windmill relics that still lingered on.

1941   The Smith–Putnam windmill generated its first power.

1945   The Smith–Putnam wind generator had operated intermittently during the war, but a main bearing had failed and with wartime restrictions, it took two years to replace. Operations began again, but in March one sail broke off at the root and operation ceased. A detailed assessment of the installation, performance and costs showed that production units would cost $190 per installed kilowatt. Conventional oil, gas or coal power stations cost only $125, so it was concluded that the wind generation of power was uneconomic and the Smith Corporation, who spent more than $1 million on the project, abandoned it. Wind power could never generate power on a large scale or could it?

1948   The Second World War had almost completed the demise of the English windmill. There were probably no more than ten in wind, with a further forty or so reasonably complete. Of those in wind, six post and one tower mill were actually operating commercially in Suffolk. Interest, however, had strengthened for the restoration of this unique machine, so detailed logging of what actually remained continued.

In America, rural electrification was being completed and the farm windmill was no longer needed there, but there was still a need in other remote parts of the world like South America, India, and Australia.

1949   The unique Bozeat post mill in Northamptonshire was wrecked in a storm.

1950   King coal's dominance of energy supply was eroding rapidly! At £2 per tonne it was still cheap, but the convenience of oil and gas had got these a big share of the market. Coal had 60 per cent, oil 25 per cent and natural gas 10 per cent, so who wanted wind power?

1953   The last wind pump at Ashtree Farm, Norfolk, stopped working.

1954   Only fifteen to twenty windmills were left in England.

1957   Thirty windmills were restored and capable of grinding flour, and others were being identified for similar restoration, but the last working corn mill at Billingford in Norfolk shut down.

1970   A hundred windmills or so had been – or were being – restored, with most of them able to mill as of old. More and more enthusiasts were joining the band of dedicated people intent on getting more windmills in wind, so that future generations could enjoy the splendour of the English windmill. But was this all that the ever blowing wind and limitless source of energy could be used for?

1974   The first oil crisis happened and became a world energy crisis! The large and increasing dependence on oil for energy became politically unacceptable, but the developed alternatives of coal and nuclear power have serious environmental problems. Clean renewable energy was needed and the wind, with data from the Smith–Putnam tests and some small wind units already operating, was a prime candidate. Modern science and technology, and more importantly Government investment, became available to quantify the power in the wind and exploit its potential as a source of power and reduce the dependence on oil.

Wind speed varies enormously (say from 0 to 120mph) in seconds, minutes, days and years and direction changes are full circle over the same time. For the best location of windmills the wind patterns need to be known, so large studies on them began in the western world.

1975   100kW, two-bladed windmills were being installed in the United States. The generator was located at the top of a 30m tower with 37.5m-diameter sails rotating at 40rpm driving a speed up gearbox to get 1,800rpm on the generator. This is the speed required to generate 60 cycle AC which is used there. Sophisticated sail angle control kept the rotor rpm constant in wind speeds above 18mph and a powered gear drive kept the sails into the wind. The unit was sold for $5,000.

The mill has, in fact, come full circle; in 1200 the original design was a box sat and turning on a short oak post – the post mill. In 1975, the generator sits and turns on a very tall steel tower – the post. The design concept is the same.

1980   55kW wind turbines were being installed in Denmark. One, two and three sail types were tried, and eventually the three sail has become the practical optimum on cost and stability in varying and veering winds. The sails must be light and rigid. Initially steel and aluminium were tried but they suffered from corrosion and fatigue (a metallurgical term for breaking due to repeated bending), so glass-reinforced plastics (GRP) have been progressively introduced to avoid these problems and now dominate as sail material. The sails twist from 13 degrees at the hub to 0 degrees at the tip. The wind turbine will start generating power at a wind speed of 3m/s and reach rated power at 15m/s and continue generating up to 25m/s when the over speed control will feather the blades (i.e. turn them out of the wind) and bring the turbine to a halt.

Power can be generated in varying wind speeds via a gearbox turning at 1,500rpm for 4 pole 50 cycle AC that is the norm for Europe. Such power can be fed through invertors or transformers into the local grid.

1991   The first wind farm was installed in England. It was at Delabole in Cornwall with a capacity of 4MW from ten turbines. These turbines had 15m-diameter sails and 30m towers.

2000   In the UK alone, 115 turbines were generating 400MW, and globally, 10,000 were turning. They are all designed to operate remotely and run for twenty years with minimum of maintenance. On wind for 3,000 to

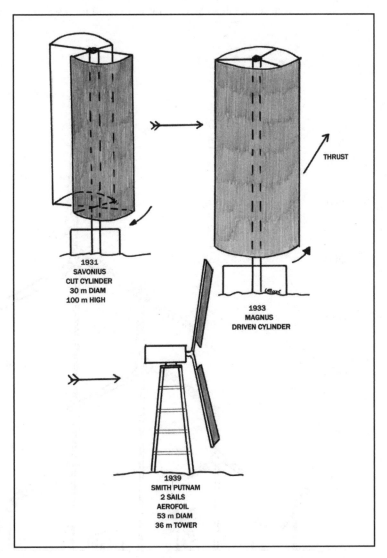

Figure 11. Experimental power windmills: Savonius, Magnus, Smith–Putnam.

4,000 hours per year, they have an operational life of 40-60,000 hours. (Compare this with the life expectancy of a car engine of about 6,000 hours.) This increase in capacity has been achieved by developing the size of the turbine so that the capital and running costs per unit of electricity have decreased significantly.

2003　　Modern turbine size has continued to increase and the size compared with houses and the traditional windmill is illustrated in Figure 12. In addition, the following table clearly shows the rapid increase in power generated.

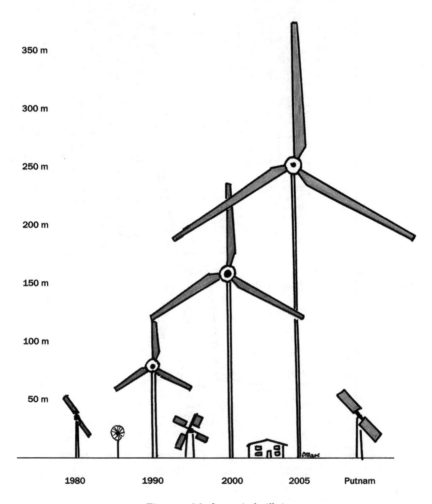

Figure 12. Modern windmill size.

TURBINE SIZE AND CAPACITY

Year	Sail diameter m	Height m	Output kW
1980	15	30	50
1985	20	40	100
1990	40	80	500
1995	50	100	600
2000	80	160	2,000
2005	125	250	5,000
American farm windmill	5	20	0.5
English windmill	20	12	16
Smith–Putnam 1939	53	33	1,250

The comparison with the older windmills shows how things have changed and how forward looking the Smith–Putnam windmill of 1939, was. Tower height has increased because wind speed increases with height following the 1 in 7 rule, i.e. double the tower height and the wind speed increases by 10 per cent. Since power increases with the cube of the wind speed this modest increase in wind speed causes at 33 per cent increase in power generated. Figure 12 shows how dramatic these increases are. In a 15mph wind, the sedate English windmill will feel the effects of 2 tonnes of air per second, while the sophisticated giant of the turbine deals with 175 tonnes of air per second so it really has to be robust.

What is the future? More English windmills will be restored and kept going so that people can enjoy the views that include them and wonder at the ability of long-gone craftsmen to build and operate what was a vital piece of machinery in their daily lives. Without modern science, they developed sails, gears and the grinding stones using wood and hand shaping until cast iron was introduced, to make flour for their daily bread.

The further development of the wind turbine is driven by political and economic factors. Currently 40,000MW have been installed round the world and growth is forecast to continue at about 30 per cent per year for some time to come. In the EU alone, by 2012 capacity will reach 75,000MW, which is 5.5 per cent of demand, and by 2020 it will be 12 per cent of demand or 200,000MW. Who knows how big the turbines will be by that time? Or what novel technology they will incorporate? What ever it is, it will be the result of the belief of individuals like Lee, Smeaton, Meikle, Cubitt, Perry, Jacobs and Putnam in their ideas that will make the next breakthrough possible and happen.

CHAPTER 3

Post Mills

T he turning of mill stones to produce flour was well developed through beast and watermills by the late twelfth century, but in the areas where water power was not suitable there was a growing desire for some other form of power. The strength of the wind was readily apparent, but how could it be harnessed? The concept of the post mill to use the wind was an extraordinary novel leap from under-driven stones in watermills and vertical-shafted fixed windmills to over-driven stones in a structure that could be turned to face the wind from whichever direction it blew. The idea could have come from the Middle East via the crusaders but it is more likely that an educated (and wealthy) practical Englishman, with a need for grinding corn where watermill operation was not possible, conceived the post mill in the early part of the twelfth century.

Buildings always have robust foundations and the structure on top rarely exceeds the area of the foundations, and the weight of the base is greater than the upper stories. The post mill is completely the opposite! It has a small diameter post holding up a hut that has a floor area ten to fifteen times the area of the post. Furthermore, the major weight – i.e. the stones – has to be several feet above ground level. The sails have to be fixed to a horizontal shaft in the top of the house, adding more weight at a high level, with most of that weight at the front end giving a balancing problem. The last and biggest problem is to control the rate of rotation of the sails in varying winds, and to stop the mill when the winds become too strong. The principal material of use was wood – several varieties – the strongest of which was good old English oak, and there was plenty of that. Iron was available from the blacksmith but, because of its cost, was used for a very few critical parts like the spindle and support for the runner stone.

The earliest mill could have been based on a growing tree with the trunk trimmed to be the post (Figure 3). The roots would be sufficient to keep the mill upright until rot weakened them. The next step was to have a cut oak post buried in the ground, but this would loosen due to the rocking caused by the buffeting wind particularly in wet ground. Over the following hundred years, several ideas were tried to stabilise the post and keep it vertical:

Stone-lined pits for full tree trunks up to 3ft diameter to be jammed in.
3 or 4 props from ground to just below the floor.
Horizontal beams laid on the ground and linked to the post and foot of the prop.
Stone foundations for the cross beams.
Earth mounds rammed round the cross beams and props.

Due to these problems, the new machines were not cheap to build or maintain, and only the land-owning aristocracy and the Church could afford to pay for them. In the late twelfth century, a post mill cost about £10, of which the pair of mill stones would be £3 to £4. A new stone after thirty or forty years would be about £2. Compare this with the annual wage of a labourer of about 1*d* a day or £1 5*s* a year. A few estate accounts from this same period show the average repair costs to be the same, i.e. £1 5*s* a year, which includes the stones, sail cloth, iron, oak for boards and beams and tallow for greasing.

Rot and rocking, however, continued to cause regular instability problems needing frequent attention and repacking to keep the mill in operation. It is likely that replacement of props and beams was a regular occurrence, with the main post replaced after fifty or sixty years. The cause of these problems was that the post not only carried the load, but also had to be truly vertical to allow the hut to turn on the pivot. These two engineering functions had to be separated and solved, or the post mill would remain a small exasperating machine to operate.

By the late fifteenth or early sixteenth century, the solution for instability had evolved by trial and error from a combination of the five basic systems already being used, plus one entirely new idea. The post no longer went into the ground! The resulting system was both ingenious and sophisticated. The main post of oak was up 5.5m long, 750mm square at the bottom, tapering to 600mm diameter at the top with a weight of 1.5 tonnes. The beams, now referred to as cross trees, were 300 by 500mm with the longer side vertical. At 6.6m long, they weighed about 0.5 tonnes each. The multi-tonne load of the house or buck was still supported by the top of the post, but this load was transferred to the props or quarter bars (250mm square, 3.5m long, weighing 0.25 tonnes) and then to the cross trees. The quarter bars were mortised into the post just under the lower floor and locked into the cross trees by 'birds mouth joints' (Figure 13), held by iron bands. Some used pins or bolts to repair or strengthen an aged joint. There were usually two cross trees and four quarter bars but at Costock, Notts, Bledlow Ridge and Stokenchuch, Bucks, Chinnor, Oxfordshire and Moreton, Essex, there were six quarter bars and three cross trees. The cross trees were not jointed into the main post nor to each other. They were continuous from opposite quarter bars and rested on top of each other so their ends under the quarter bars had to have foundations of different heights. This was achieved by using stone pillars or piers that also raised the cross trees above ground level. This allowed rain to run off the trestles and let them remain dry for most of the time and so reduce the rotting problem. The weight of the mill had therefore been transferred from the post to the stone piers (Figure 13, Colour 1 and 25).

The post was also fixed in the vertical position and this was done by cutting slots in the bottom of the post, leaving legs or horns to straddle the cross trees so that the post could not move laterally in any direction. Critically, the post did not rest on the uppermost cross tree. A slot was cut so that there was a gap of 12 to 25mm between the bottom of the post and the top of the cross tree so that no load was transferred to the cross trees at the centre point, and with the skill of the carpenters, the post was locked in the true vertical position. This method of construction became standard and was used right into the nineteenth century (Figure 14).

Developments were not, however, restricted to the stability problem. Initially, the wind shaft was horizontal, holding simple cloth sails only 2m long and sweeping to within 0.5m of the ground. The post is set nearer to the sail end of the buck or house so that the stones are at the back to act as a counterbalance for the weight of the sails. One pair of stones, probably 3 to 5ft diameter, would be driven directly from the big wheel. By 1300, as the stability improved, larger cloth sails that could be full or partially furled to compensate for varying wind speed were in common use. The sails almost always turned 'against the sun' i.e. anticlockwise. This came about because the miller had to climb up the lattice to furl the sails. Most people, including the millers, are right handed, so he needed his right hand to deal with the sail cloth and the ropes, and he held on with his left. The sails are furled round the sails stocks which must be on the right with lattice on the left. The lattice is on the trailing edge of the sail so the sails must rotate in an anticlockwise direction.

The wind shaft was initially horizontal when the buck was supported on a single post, but the solution to the stability problem was the quarter bars, cross trees and piers, the extremities of which protruded beyond the front of the buck. With a horizontal wind shaft the sail tips would have struck the ends of the quarter bars so inclining it at 8 to 15 degrees, moved the sail tips clear of the mill supports. With stability no longer a problem, the mills were built higher to make longer sails possible and use the stronger winds to get more power. With this extra power, a second wheel, the tail wheel, was added to the wind shaft to drive a second set of stones. The gearing between the big wheels and the wallowers that drove the stones was about 4.5–5.5 to 1, which, with 6m-long sails rotating at 15 to 20rpm gave stone speeds of 70 to 110rpm to produce a large volume of good flour. In later years, gearing as in tower mills, particularly for higher post mills with longer sails, made it possible to get stone speeds of 100 to 120rpm.

All these improvements increased the already large weight of the buck which had to turn on a simple pivot. The original type had a 200mm-diameter pintle or pivot made of oak, or a gudgeon pin made of wrought iron. Later, after the Industrial Revolution, a cast-iron cap with an integral 150mm-diameter iron pintle was also used (Figure 15). On the pintle turned the crown tree, a massive beam 375mm wide by 400mm deep which ran the full width of the buck at right angles to the wind shaft. This beam carries the multi-tonne load of the buck and its contents, and bucks were about 4.8m long by 3m wide while the largest was 6 by 3.5m. The crown tree was at the vertical mid-point of the buck with the stone floor resting on it and a lower floor used for grading, sack filling and storage. The stairs to the ground lead from the rear of the lower floor and were as wide as 2m, and through them was the long tail post for turning the mill into the wind. The weight of the stairs helped to counterbalance the weight of the sails, buck and stones during turning, and when they rested on the ground, locked the buck into position. When the miller wanted to turn the buck, he used a secondary lever called the 'talthur' (Figure 14), to lift the stairs off the ground so the mill could be turned into the wind. With full bins of grain, the larger post mill could weigh 20 to 25 tonnes and all this was balanced on the pintle and pushed round by the miller.

Space was at a premium to keep grain and flour under cover and have grading and cleaning machines in the two-storey mill, so the enterprising miller added side panniers and rear porches to get the much-needed space (Figure 16). To get more power for grinding, even higher mills were needed. To achieve this, the post was increased in length and the brick piers were built up to 3.5m in height. Underneath these piers was very useful space, partially protected from the weather by the buck. To complete this protection round houses were built round the piers, providing not only storage but also protection of the oak mill supports from rain (Figure 16 and Colour 2). Often the walls of the round house incorporated the piers, but some were built with walls outside the piers to get more protected space. The first record of a round house was at Foston Mill in 1760 which had a buck of 5.1 by 3.3 m. Since round houses could be added without interference to the operation of the mill, they were soon built under a lot of post mills. Two doors were needed at opposite ends of a diameter to ensure access the round house without dodging sails when one door was obstructed by the low sweeping sails which were often only 600 to 750mm from the ground.

Skirts or petticoats were fixed to the bottom of the front face of the buck to deflect rain away from the centre of the roof. Some fastened tracks to the top of the wall and rollers to the bottom of the buck which aided stability during turning in high winds. Cart wheels were attached to the end of the tail post and the ground levelled to help the smooth and easier turning of the buck which, when it was well balanced, could be easily turned by the miller or his helper. Unfortunately, it was not always that easy. Winches on the tail pole hauled the mill round to posts driven into the ground and sometimes a horse was harnessed to the tail post to provide the power, but horses occasionally bolted and the mill could make few complete revolutions before the horse was calmed so this was not common practice. In some cases, a more elaborate wheeled structure was added to the tail post and a fantail used to drive the wheels for automatic turning into the wind, but the strong back of the miller remained the common (if not popular) way of turning the mill.

The larger post mills and those with round houses were permanently sited, and since the main structure was now protected from the weather, rotting was a lesser problem, so mill life before major repairs were needed was extended to eighty to a hundred years. The smaller post mills were regarded as moveable, even though the task would have been considerable. They would have been moved to get a better and windier site, or when they were sold to a new owner. The mill was lowered onto rollers under the cross trees and dragged to its new site. It is recorded that a team of eighty-six oxen were used to move a mill two miles near Brighton, and in London one was moved by barge. In the Nottingham Forest area there were thirteen post mills, one of which was moved so often that it was known locally as 'Roving Molly'. The same name was used in Lincolnshire, so it may be that any mill that was moved was called a Roving Molly. The post mill was also known as the German mill, particularly in the mid-1800s.

The walls and roof of the buck are always exposed to the weather. Thatched roofs and wattle and daub walls were cheap but not durable, though they did persist for a century or more. Boarding was soon the preferred alternative, sometimes using two layers to get real weather proofing of the mill. Bare wood was replaced as it rotted until white lead paints and tars became available for protection of the wood – the one used being the cheapest available locally. In the nineteenth century, a few mills were sheeted with steel or felt and then painted. In the Mediterranean region, more basic post mills were made. The crude wooden buck just cleared the ground and rotated

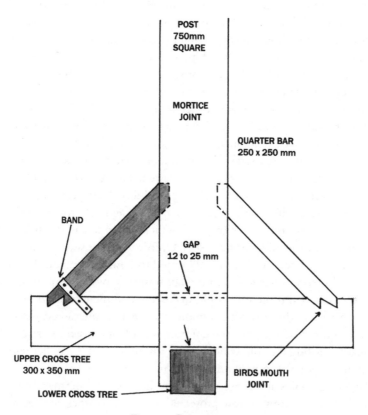

Figure 13. Post supports.

from a point very near to the front. A small single pair of stones was driven in the one-storey buck. Being in a hot dry climate, the wall and roof boards were left bare.

Up to the late eighteenth century, all post mills had four common sails, but after Miekle invented the spring sail, some began to convert two of them to this new system retaining two common, while some converted to all spring. When Cubitt's patented sails were invented in the early nineteenth century, the second mill to be equipped with them was the Horning post mill. By the 1880s, when the total number of mills was estimated at 10,000, there were probably about 2,000 post mills; many would have a combination of spring and common sails, some fitted patent sails, more had all spring and a lot retained all common sails. Almost all had four sails rotating anticlockwise driving two pairs of stones – one pair from the brake wheel and one pair from the tail wheel. Between 30 per cent and 50 per cent would have round houses.

1 Brake Wheel
2 Tail Wheel
3 Windshaft
4 Neck Bearing
5 Breast Beam
6 Tail Bearing
7 Tail Beam
8 Stone Nut
9 Runner Stone
10 Bed Stone
11 Side Girt
12 Crown Tree & Pivot
13 Post
14 Quarter Bar
15 Cross Tree
16 Brick Pier
17 Tail Pole
18 Talhur
19 Steps
20 Grain
21 Meal

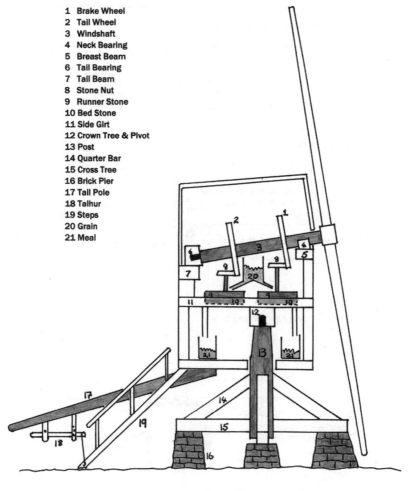

Figure 14. Post mill.

Specific details of unusual and individual post mills are given below:

Bourn, Cambridgeshire (Colour 1).

Wrawby, Lincs, has a round house and two common and two spring sails (Colour 2).

Hermonceaux, East Sussex, was 15m to the ridge and had patent sails controlled by the Hammond system.

Hogg Hill had a roof-mounted fantail.

Ashcombe, Sussex, had six double patent sails mounted on the wind shaft by a three-way poll canister and a neck bearing of glass.

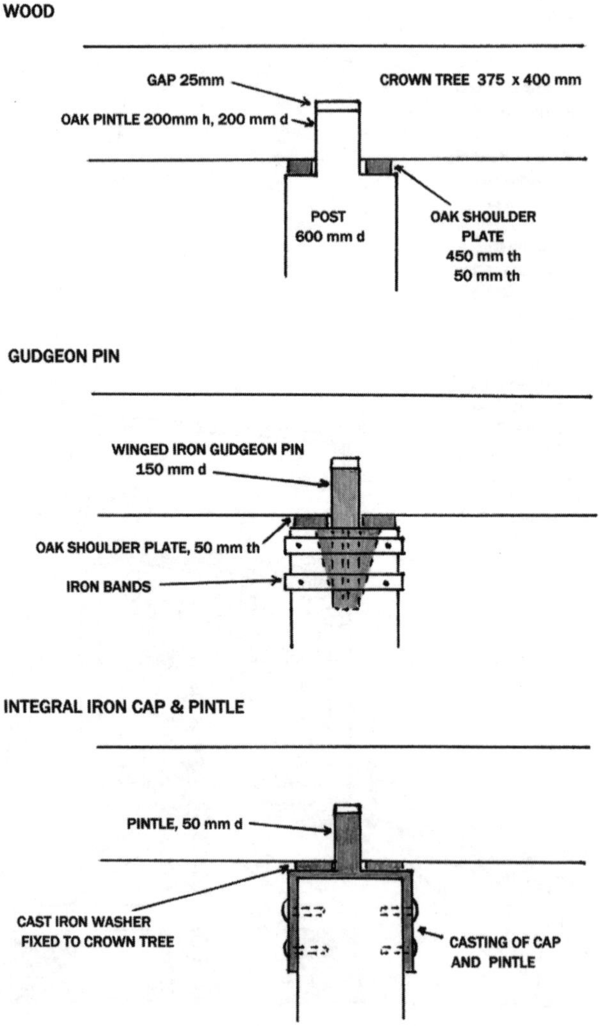

WOOD

GAP 25mm

OAK PINTLE 200mm h, 200 mm d

CROWN TREE 375 x 400 mm

POST
600 mm d

OAK SHOULDER
PLATE
450 mm th
50 mm th

GUDGEON PIN

WINGED IRON GUDGEON PIN
150 mm d

OAK SHOULDER PLATE, 50 mm th

IRON BANDS

INTEGRAL IRON CAP & PINTLE

PINTLE, 50 mm d

CAST IRON WASHER
FIXED TO CROWN TREE

CASTING OF CAP
AND PINTLE

Figure 15. Detail of pivots.

Lower Dean, Bedfordshire, had four patent sails rotating clockwise and both a tail pole and fantail.

South Skirlaugh, Yorks, had, in 1852, three pairs of stones – two pairs driven from the brake wheel and one pair from the tail wheel.

Wetwang, Yorks had four patent sails controlled from a rear platform built on the steps.

Pitstone, Bucks, may be the oldest surviving post mill, as there is a date carved on the lower side girt of 1627. It also has two pairs of stones side by side driven from below via a spur wheel.

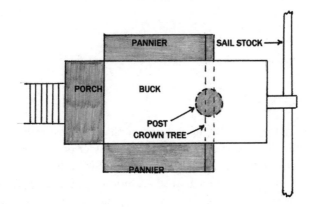

ROUND HOUSE WITH CURB

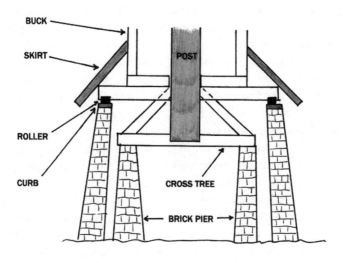

Figure 16. Post mill plan and roundhouse.

CHAPTER 4

Smock Mills

The desire for more height and storage than the post mill could provide, and the difficulty of turning the larger and heavier post mills, meant a complete rethink of how mills could be built. In the fifteenth century, it was the Dutch that did this. It was not a gradual evolutionary process but a radical change in mill design. The buck would be fixed and only the top section would rotate. A fixed body was stable and could be built higher to get stronger winds and more power for milling or water pumping, which would be the prime concern of the Dutch. The top section, called the cap, held the wind shaft that carried the sails and the big main driving wheel, and only these had to be turned to face the wind. This weight was considerably less than that of the fully loaded buck of the post mill, and the strong fixed walls supported a wide circular curb on which the cap rested and turned. (Compare this with the single oak pivot on which the post mill turned.) Because the weight of the cap was on a large diameter, the cap was very stable, and because the curb was broad, its load was spread over a large area of sliding surface, making it relatively easy to push the cap round. This was just as well, for the turning pole length had to be increased and was inclined at a steep angle. The pole, or double pole, was joined into the main support beams that carried the bearings of the wind shaft. The stability and turning problem had thus been solved.

One set of stones located centrally in the body would still be driven by a lantern wallower, as the main shaft from the wallower would be at the centre of rotation of the cap. However, at an early stage, the next radical invention was made. It must have been realised that, by removing the single pair of stones, the central shaft could be extended down to the lower floors of the mill and, with additional gearing, more stones could be turned. A large wooden gear wheel on this shaft could drive two or more pinions that themselves drove the stones and eventually lead to mills with up to four pairs of stones. The large gear became known as the spur wheel and the stone drive wheels as stone nuts. The simple drive in the post mill of big wheel to wallower gave a speed–up ratio of 4.5 to 5.5 to 1 and a stone speed of say 70 to 110rpm. With a two-step gear train of big wheel to wallower and then spur to stone nut, speed ratios of 6 to 12, higher stone speeds of about 110 to 130rpm became possible at the lower sail speeds of longer and wider sails. With these speeds and three pairs of stones, the output of a finer flour for better bread increased significantly.

The wooden mill walls are built on cant oak posts of 225mm², with the lower end buried in the ground. The mills were usually octagonal, but some had ten or even twelve sides with a maximum of four floors, as more than this in wood alone would flex in high winds and with age. Figures 6 and 17 show the section through the mill. To get more height, a brick base with two or three floors was constructed, and the wooden cant posts fixed to this, but the practical maximum was six floors with a base diameter of 5 to 7.2m. These mills were often painted white and their resemblance to the agricultural workers' smock is said to have given them their English name. There were, however, a lot in black, as Colour 3 and 4 show. The earliest record of such a mill is Green's Mill, erected in Buckinghamshire in 1650, and the highest smock mill was Union Mill at Cranbrook at 23m high.

Flexing of the wooden structure remained the major problem. It caused movement of the sill blocks to which the curb was fixed, making turning difficult, and loosening of main structural joints, which caused the weather boards to crack and let in the rain, causing rotting of them and main beams. Maintenance as the mill aged became a serious and expensive problem, so when major repairs were needed some smock mills were rebuilt as brick tower mills when the octagonal shape usually changed to a round tower.

CHAPTER 5

Tower Mills

Perhaps the earliest tower mills were those original ones in Seistan in the Afghan-Iranian hills. They were, after all, squat towers built from the most durable of natural materials – stone. In latter centuries, and to more recent times in the Mediterranean region, round-fronted short stone towers were used to point the sails at the prevailing wind. The remains of such towers can still be found in the Ambelos Pass at Losssithi in Crete. Neither of these types of mill, however, had turnable tops, and the first tower mill was recorded at Dover Castle in 1295. For some reason, no more seem to have been built, and it was not until the sixteenth century that tower mills appeared again in the Flemish region of Europe – and this version clearly evolved from the smock mill. After a hundred years of flexing wooden structures, the change to a permanent, rot-proof and stable tower of stone or early brick was inevitable and logical. They usually had four floors, though Southdown Mill at Great Yarmouth had twelve. Towers were round, but where a smock mill had been rebuilt as a tower mill the octagonal shape was sometimes retained. Two doors at opposite sides were the norm, and windows were set two on each floor vertically above each other or offset at right angles on alternate floors. The walls were at least 450mm thick, and as brickwork weighs 1.8 tonnes per cubic metre the towers weighed several hundred tonnes. The tower of the Wendover mill had 900mm-thick 25m-high walls, giving a ground loading of 20 tonnes per square metre, so substantial foundations were needed. Because of the inclined wind shaft, almost all towers had sloping sides whose angle was called the batter. A simple triangular wooden frame of the desired batter, usually between 8:1 and 10:1, with a plumb line attached to the vertical side, was used in building to maintain this constant angle. The area of the floors increased from top to base, giving both a stable structure and more working space than a parallel-sided tower.

With no scaffolding to hand, the mill was built from the inside. Walls up to floor one could be built from trestles on the ground, and then floor one was installed and used for the trestles to build floor two and so on until the required height for the curb was reached. Examples of mill towers are given below which shows their wide range of dimensions:

	Base	Curb height diameter	Curb diameter
	m	m	m
Heage, Derbys	7.2	8	6
Hewitt's, Lincs	6.4	12	4
Denver, Norfolk	7.8	18	6
Southdown, Gt Yarmouth	14	30.5	5

The more usual tower had four floors, each of 2.5 to 3m (12m high) with sails of 7.5 to 10.5m long, sweeping to about door height. Where mills exceeded four floors, sails remained at no more than 13m, as this is the practical limit for wooden sails. A balcony, therefore, was usually attached four floors down from the top for operation of the shutter and brake chains, though some six-floor mills had no balcony so the brake and shutter control chains extend all the way to the ground. The standard layout from the top down is shown in Figure 17 and listed below:

Dust floor grain sacks are taken to this floor using an endless or single chain hoist driven from the wallower which is sometimes called the crown wheel.

Bin floor the grain sacks are emptied through holes in the dust floor into bins on this floor to provide ample capacity of grain for milling.

Stone floor at this level are two to four pairs of stones for milling the grain to flour.

Meal floor the ground meal can be bagged directly from the stones, binned or fed into sieving machines for cleaning and grading of the flour.

In a three-storey mill, the dust and bin floors are combined, and in three and four-storey mills the meal floor has to be used for bagging and storage of flour prior to delivery to customers and can be a very crowded place. Additional floors made working, storage and delivery planning much better and gave overall higher output for the miller. This apparently simple layout is, from an engineering point of view, superb. Grain sacks are hoisted by wind power to the top of the mill and gravity is the force to feed each stage of the milling process, resulting in sacks of flour on the elevated stage outside the door for easy loading onto the horse-drawn carts.

For weather protection, a few mills did have higher quality facing bricks to keep out the rain, but the most common protection was white lead paint, lime render or tar, and there were some interesting regional variations in their use. Eastern mills, particularly in Lincolnshire, were tared black with white sails and caps, whereas in Lancashire the mills had white towers with black sails and caps.

Since tower mills were round brick structures and smock mills were wood and angular they were initially regarded as distinctly separate types, as they are now. In the eighteenth and nineteenth centuries they were not. There were brick smock mills and wooden tower mills! Thrumpton Mill in Retford was sold as a smock mill in 1841 and described as a tower mill in its sale in 1880. In part, this interchangability of terms was due to the rebuilding of wooden smock mills as brick tower mills, particularly where the octagonal or hexagonal shape of the original smock mill had been retained, and to add more confusion, for a time in the mid-1800s, tower mills

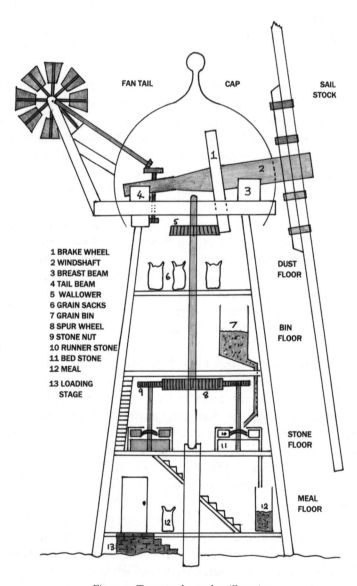

Figure 17. Tower and smock mill section.

were also known as Dutch mills. Fortunately, today, the smock mill is the angular wooden mill and the round brick mill is the tower.

In total, they reached about 12,000 out of 15,000 mills operating in the 1840s, and almost every one of them had four patent sails rotating anticlockwise. The estimate of mills with five, six or eight sails is between sixty and 100, with fewer than ten having eight sails. The number of floors ranged from three to twelve, but about half had four floors and a further quarter had five floors.

Restored tower mills, not only for corn milling but also for water pumping, can now be seen in many places in the country. These are listed below along with some that have more unusual features:

Hewitt's, Heapham, Lincs. *Colour 5.*
North Leverton, near East Retford, Notts. *Colour 6.*
Mount Pleasant, Kirton Lindsey, Lincs. *Colour 7.*
Skidby, near Hull, Yorks. *Colour 8.*
Burgh Le Marsh with five clockwise-rotating sails, Lincs. *Colour 9.*
Heckington with eight sails, Lincs. *Colour 10.*
Ellis with three floors, Lincs. *Colour 11.*
Sibsey Trader with six sails, Lincs. *Colour 12.*
Tuxford, Notts. *Colour 13.*
Waltham with six sails and, despite being a high mill, had no balcony. *Colour 14.*
Bircham, near King's Lynn, Norfolk. *Colour 15.*
Denver, near Downham Market, Norfolk. *Colour 16.*
Mediteranean-type tower mill, Feurteventura, Canaries. *Colour 17.*
Stone Cross, East Sussex, had round windows.
Clayton (Jack), East Sussex, had Hammond-patented shutter control.
Bells, Lincs, had five pairs of stones on the fourth floor. With stair entry and sack hoist
    trap doors, it must have been a very crowded floor.
Rhoads, Lincs, used Bywater-patented roller reefing sails.
Marsh Chaple, Lincs, had two patent and two roller reefing sails.
Hoyle's, Lincs, had two spring and two common sails and a tail pole for luffing.
Southdown Mill, Great Yarmouth, Norfolk, is often quoted as the highest mill, built at
    34m with twelve floors. It was also known as High Mill, Cobholm Mill and Gorleston
    Mill, but the highest mill was actually Bixley Mill, Norwich, at 42m.
Wycombe Heath, Bucks, had four common sails.
Westfield, Somerset, had only two floors, four common sails and one pair of stones.
Dunloe, Beds, had four common sails and a conical cap.
Maud Foster, Boston, Lincs, had five sails.
Much Hadham, Herts, had eight double-sided patent sails, eight floors and was 35m
    high to the cap and had four pairs of burr stones. It was the last mill built in Britain
    and only milled for three years.
Thurne Dyke wind pump, Norfolk. *Colour 18.*
Horsey wind pump, near Stalhm, Norfolk. *Colour 19.*
Wicken Fen wind pump, Cambs. *Colour 20.*

CHAPTER 6

Caps

The purpose of the cap or roof was to protect the main gearing and inside of the mill from rain. On the post mill, the whole rectangular buck turned so the roof is just a simple pitched roof like a shed. Variations on the simple pitch were a two angled pitch or a curve and a sloping ridge that was higher at the front. Thatch and sail cloth were the first materials used, followed by longer lasting overlapping boards. Once the tower and smock mills were introduced, roof design had to change.

The base of the roof or cap became circular and there was a large wind shaft rotating at the front and the cap itself had to turn full circle in either direction, so it was no easy task to solve. After the fantail was introduced, there was the added problem of a second smaller shaft and the main fantail supports protruding through the cap. Probably the first caps were based on the double angle sloping ridge developed for the post mill but, as long as they kept the weather out, its actual shape was not critical, so the whim and fancy of the owner or millwright helped to develop more regional types. The double angle pitch with sloping ridge became popular in the south of England. Further development lead to a curved ridge with curved sides where the sail end was wider than the back. The profile resembled that of an upturned boat and so became known as the boat cap, most used in East Anglia.

Starting from a circular base, the more logical shape to add is a cone or dome, and in the Mediterranean region simple conical caps became the norm. In England, the dome evolved into the ogee or onion-shaped cap which dominated in Lincolnshire. The curve staves for these ornate shapes were cut from naturally curved trunks or branches which needed only trimming to get the final profile. From an aerodynamic point of view the rounded onion shape is actually the best, as it offers the least resistance to the wind flow leaving the sails, whereas a flat wall would produce the most turbulence and reduce efficiency.

The cap types are shown in Figure 18 and Colour 1, 11, 17 and 18.

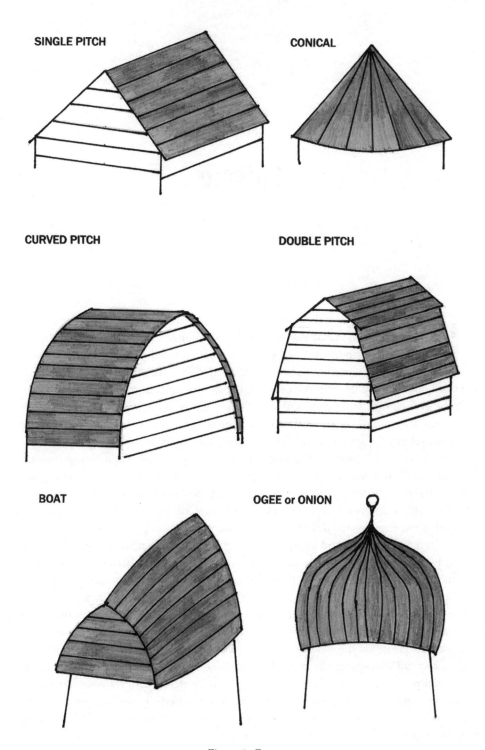

Figure 18. Caps.

CHAPTER 7

Luffing

L uffing or winding is the term used by the millers to describe the operation of turning the sails into the wind. On the post mills, the whole buck turns on the pintle at the top of the post, and for several hundred years it was done by the strength of the miller pushing on the tail pole. In the early smock and tower mills a double tail pole from the protruding cap support beams and stretching down to the ground or balcony meant it was still the miller pushing, albeit a lighter load, to keep the sails into the wind. Luffing was a critical operation, so the miller had to keep a watchful eye on the wind and luff frequently to ensure the sails faced the wind, as the mills were designed to withstand the wind's force only when facing into the wind. A mill tail winded in a strong or gale-force wind was in danger of loosing sails or even the whole wind shaft if the wind force pulled the shaft out of its bearings.

The earliest curbs on smock and tower mills were known as dead curbs because the wooden cap frame rested directly on the wooden curb planks fitted to the top of the walls. It required liberal and frequent greasing with tallow to make luffing of the sails, wind shaft and cap structure a manageable task. The Dutch initially used an inside winch to pull the cap round in their large polder mills, but this meant the operator climbing up into the top to move the sails round into the wind. They changed to the long tail poles reaching to the balcony or ground for convenience, but it still needed a lot of effort. Even when the wind was reasonable, the pressure on the curb was much greater at the sail side than the tail, and as wind speed increased lifting at the front and sliding were occasional frightening events. About the time the English began using tower mills to help luffing and increase stability, rollers were introduced into the cap frame with more at the sail side to compensate for the heavier load there, and these became known as live curbs. Centring wheels (Colour 27) were also added to the cap frame running under a lip on the curb to eliminate sliding and lifting, so ensuring the smooth and stable rotation of the cap about the central axis of the main shaft. Eventually, these live curbs completely replaced the older dead curbs. Once this was done, teeth or cogs could be added to the curb ring and a gear wheel turned by handle from the inside of the cap or by endless rope over a counter wheel to the ground, to more easily and controllably luff the sails into the wind (Figure 19).

For a few hundred years, wood was the main material available for teeth, wheels, rollers, axles and plates, with a few items of wrought iron forged by a blacksmith. Even with wheels and rollers the miller's brawn remained the motive power for luffing until, in 1745, Edmund Lee patented the fantail for automatically luffing the sails into the wind. The fantail is a six or eight-bladed wheel set at right angles to the main sails on a wooden frame at the rear of the mill. The drawing in patent No.615, dated 24 March 1745, shows the fantail frame at ground level, fixed to the tail pole and running on planks engaged by a spiked wheel. It was intended for tower mills but the patent says the system can also be used for post mills, and it was on post mills that a ground-level fantail attached to the stairs was adopted.

Edmund Lee's patent was granted in England, Wales, Berwick on Tweed and America, so it must have been a unique document that made Berwick on Tweed as important as America. In 1750, Andrew Meikle of Houstan Mill near Dunbar in Scotland invented the high-level fantail located on rear of the cap. It was fixed on extensions to the sheer trees still in the lee of the cap (Figure 20), which would make it somewhat insensitive to small changes in wind direction but a vast improvement on pulling or pushing the tail pole. The then patent law and its enforcement is unclear as to what was protected, but the Lee patent did not apply to Scotland. Meikle's invention was not only the fantail, it was a better and workable mechanical system for reliably linking the rotating fantail to turn the sails into the wind. A worm gear on the shaft from the fantail engaged a rack on the curb (probably both of wood), so there was positive coupling at a reduction of 5,000 to 1. This was also a novel concept, but Meikle did not apply for a patent even though it was the real solution to automatic luffing. Before its wide adoption, however, two other developments were needed. In 1752, Carmus designed bevel gears to get good meshing of gear wheels that were not at right angles to each other and profiled teeth to reduce wear and backlash.

Smeaton introduced cast-iron gears for Carron Mill in 1754. Such gears were cheaper than forged ones, as a reproducible tooth profile could be cast into the wheel. This new material and process were rapidly adopted, not only for more gears but also for tooth segments, shafts and bearings. By the end of the 1700s, the cap-mounted fantail with bevel gears and pinion drive onto a rack (rather than a worm) had become first choice for tower and smock mills. Post mills tended to retain the pole or use the stair-mounted fantail turning cart wheels on the ground to get the buck to turn. Colour 2 and 3 show details of the gearing and Colour 6, 11, 17 and 18 give examples of some of the cap types.

The fantail itself was 3 to 4m diameter, and about 1806 it began to be mounted on an inclined frame fastened to the sheer trees. Hessel five-sailed mill near Hull was the first to have this. It raised the fantail above the turbulence from the cap and let the fantail luff the sails at small angular changes in wind direction. By 1820, this type of fantail was standard for tower mills. When the wind is blowing directly at the sails, the fantail blades, being parallel to the wind do not turn, so the cap remains stationary. As the wind veers the fantail begins to slightly face the wind. At about 5 to 10 degrees of change in wind direction, the fantail vanes rotate (Figure 20). The gearing from fantail to curb rack is low (Colour 26) – typically in the range 1,000 to 1,200 revolutions of

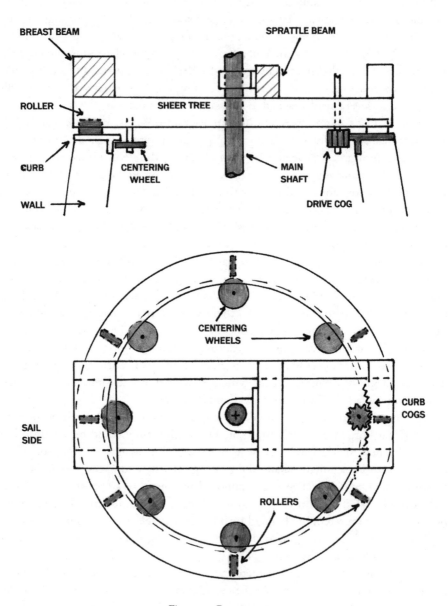

Figure 19. Cap structure.

the fantail to one complete turn of 360 degrees of the cap – so the cap turns easily even when the wind angle to the blades is low. As the vanes turn, the gearing turns the cap back into the wind and the fantail out of it, so motion stops when the main sails are realigned into the wind. The fantail will turn clockwise or anticlockwise to bring the sails back into the wind, and is left in gear even when the mill is not grinding to ensure that the sails face the wind at all times, which is the safest position.

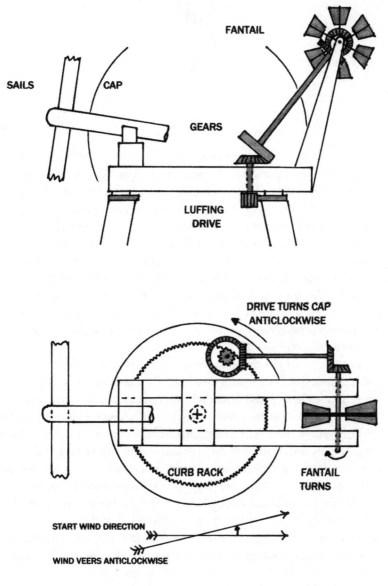

Figure 20. Fantail operation.

CHAPTER 8

Sails

The wind has driven ships for centuries and the sailors learned the best shape, size and angle of the sail to the wind to propel the ship. If a ship is at right angles to the wind, it heels over showing the power in the wind, but it does not move. As the ship is angled to the wind, it heels over less and moves forward. By varying the angle to the wind the sailor can maximise the forward speed, and if he brings the ship directly into the wind, movement ceases. As wind speed changes he can change the sail area by reefing or furling of the sails and so control the speed of the ship. Sails on a shaft behave in the same way. When flat to the wind, they feel the full force of the wind but do not turn the shaft. When angled out of the wind, the force that caused the ships to move caused the windmill sails to rotate. When the sails were head on or parallel to the wind there was no force on the sails and hence no rotation (Figure 21).

Over a century or so, the simple-vertical shafted mill with thatched sails set flat to the sail supports was combined with sailing ship know-how and an outstanding unknown genius made the wind shaft horizontal and angled the sails to the wind. The sails on the first post mills of the twelfth century in England were simple but effective. These primitive sails were short, rectangular, centrally supported and probably bound to the arms (Figure 22). Rarely, illustrations show the sails to be broader at the tip than the hub, and this was probably an early attempt to get more power when longer sails were not possible. Some sketches show the canvas permanently tied to the frame, making reefing impossible and controllable only by turning them partially out of the wind. A little later, pictures show the canvas fixed at the hub end and passed under and over the bars at right angles to the arm, and such sails could be reefed to vary the surface area according to the wind strength (Figure 23). By changing the angle and seeing what happened, an angle of 20 to 30 degrees would be found and the sails would have rotated steadily, wind permitting, usually in an anticlockwise direction. There were four sails from the very beginning, and this medieval design of sail persisted for a hundred years or so.

The figures to illustrate these changes convey an apparently clear cut progression from one to another, but nothing could be further from the truth. There would be an overlap of tens of years in the declining use of the primitive sail and the take up of the medieval, which would start at individual mills remote from one another. The same applies to the change to the common sail (Figure 24 and Colour 28), which still persists, and the adoption of spring and patent sails (Figure 25 and Colour 29). Change usually occurred when a new mill was built or when sails needed replacing.

Sail length was limited by the availability of straight timber that could be obtained which was no more than 12m. When such a stock is mortised through the end of the wind shaft there is 6m per side, restricting the individual sail to 5m long. Probably by the fourteenth century the introduction of the sail whip made sails double this length possible. To the 6m of stock on each side of the shaft another lighter beam (called the whip) up to 12m long could be strapped. With 5m of overlap, the whip was very firmly held, and the sail bars were mortised into the whip and not the stock. These better carpentry techniques also meant that the sail bars could be accurately angled and the optimisation of sail pitch to 15 to 18 degrees probably occurred at this time. These sails could also be made independently of the stock and attached to it once complete. With the double-sided medieval sail, eight cloths needed to be adjusted so it would be logical to move the whole sail to the trailing side of the sail and so have only four cloths to deal with. From the single-sided medieval sail, it is only another short evolutionary step to put the sailcloth on the face of the frame, and such sails became known as the common sail. The framework for these sails had bars both parallel to the arm and at right angles to it, making a very sturdy frame that was attached only to the trailing edge of the whip. The canvas was fastened on the face of the frame and fixed to the whips, but at the heel end the sail had rings or loops to let the sail move on the sail bars. When stopped, the sail was rolled up and tied to the whip. They must have been easier to reef than the under and over medieval sail, and since they offered a smooth profile to the wind the miller would find that he had more power for milling for the same area of sail. Even with these advantages, it would be tens of years before the last medieval sail was replaced.

Depending on the strength of the wind, the miller would unfurl the sail roll. In a light wind, the sail would be set full. As the wind strengthened, the sail would be furled up to 1st reef – about ¾ of full sail area; then dagger point – about ½ sail, and finally to sword point – about ¼ sail. To do this, the mill had to be stopped with a pair of sails vertical, so that the lower sail could be reefed by the miller pulling on ropes or clambering on the sail frame, much as sailors had to do to set the sails on the masts and rigging of a ship. The mill, therefore, had to be stopped four times to reset each sail so flour production would be stopped for 20 to 30 minutes. The miller would therefore use his experience to set the sails and reduce the number of times he had to reef. He would also decide on the number of stones he could use in a given wind. Control of stone and wind shaft speed was therefore crude and milling had to be carried out over a wide range of wind speeds, from 8 to 40mph. This means that the grinding process was also done over a wide range of speeds, from 80 to 120rpm, not only varying production rate but, more importantly, the fineness and quality of the flour.

While reefing can cope with varying winds, the common canvas sail cannot deal with sudden gusts, which caused a rapid increase of stone speed for a short time, and in extreme cases damage to sail canvas or frame. The sail was, however, reasonably efficient, light and cheap to replace, but required the art, experience and skill of the miller to set. He could also signal to his customers with the sails. Sails stationary in the diagonal position (St Andrew's Cross) indicated that the mill had shut down for some time, as this position of the sails was the most stable because the sail tips are at their lowest. Sails stopped with one pair of sails vertical (St George's Cross) meant the mill was stopped for a short time only. A more subtle variation of the St George's Cross was that 'good news' was indicated if the top sail was just before the vertical and 'bad news' if it was just after.

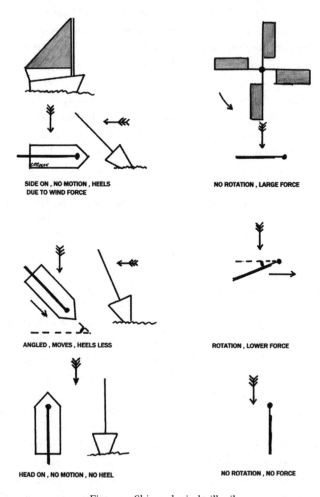

Figure 21. Ship and windmill sails.

When post mills got bigger and sails longer, two pairs of stones could be used, one in the front and one in the tail. Rarely, using spur and stone nuts, two pairs could be accommodated in the front while retaining the tail pair. In tower and smock mills with four or more floors and three or four pairs of stones as standard and occasionally five or six, the power needed from the wind was made possible by the addition of the whip, which was strapped and bolted to the sail stock. The whip was made of straight grained pitch pine or memel, 225 to 300mm wide and 300 to 350mm thick at the poll or shaft end, tapering to 150mm² at the tip. The one-angled common sail had reached the limits of power extraction from the wind so changes were needed, but it remained the norm in England. The Dutch, however, were trying new ideas to get more power from the wind, particularly at low wind speeds, because water needed to be pumped all the time to stop land from flooding. Around the beginning of the seventeenth century, mills were being built with sail twist, i.e. the angle to the wind at the hub was 20 to 25 degrees, lessening to 1 to 5 degrees at the tip (Figure 26 and Colour 30). Also, about the same time, instead of using a single-sided sail, the sail was moved on the arm to have most of the sail still on the trailing edge and a part on the leading edge. The ratio of the widths became known as arm divide, and by trial and error settled at about 1 to 3, with the narrower part on the leading edge. These Dutch developments increased the efficiency of the sail so that they would turn in low wind speeds. This was very important for the water pumping drainage mills as they would operate more of the time than the less efficient common sail in England, where it would be a further 200 years before sail twist was widely used.

The English and Dutch sails were made of stout timbers and heavy. A set of four would weigh 3 to 4 tonnes, and this has a beneficial effect on sail speed in varying winds since the momentum of the heavy load took some time to speed up or slow down as wind speed changed. To some extent, a short gust would be unnoticed

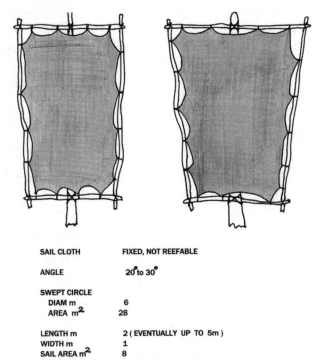

SAIL CLOTH	FIXED, NOT REEFABLE
ANGLE	20° to 30°
SWEPT CIRCLE	
DIAM m	6
AREA m²	28
LENGTH m	2 (EVENTUALLY UP TO 5m)
WIDTH m	1
SAIL AREA m²	8
% of CIRCLE	28
GUST PROTECTION	NONE

Figure 22. Sail evolution – primitive.

by this system. Mills in the Mediterranean region and the Canaries were very different and influenced by the lack of locally available, substantial, long, straight oak or pine for sail stocks. Here the cloth sails were triangular like the jib sail on a yacht. Four to twelve round slender spokes 5 to 6m long were bound or clamped to the wind shaft and their tips connected by guy ropes to an extension of the wind shaft to give support. Other ropes – tip to tip – gave circumferential stability (Figure 27). The triangular sails were wound round the spokes with the point roped to the following tip. In the wind they take up an aerofoil shape, giving good power at low speed. In high winds, the rotation speed increases and tends to disrupt the aerofoil shape so less power is taken from the wind making speed self-controlling to a degree and reefing an infrequent task. When needed, the mill would have to be stopped four to twelve

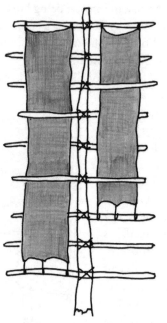

SAIL CLOTH	**UNDER AND OVER SAIL BARS, REEFABLE**
ANGLE	20° to 30°
SWEPT CIRCLE	
DIAM m	12
AREA m²	113
LENGTH m	5
WIDTH m	1.5
SAIL AREA m²	30
% of CIRCLE	27
GUST PROTECTION	**NONE**

Figure 23. Sail evolution – medieval.

times to wrap or unwrap and refasten each sail – a lengthy task, even though the sails swept within 1m of the ground and reached only halfway up the spoke. Their towers were also very different from the English type – they were squat, built of stone with parallel sides and had a conical thatched top (Colour 17).

Wind strengths described in words mean different things to different people, and while light, strong and gale mean increasing winds, they are unquantifiable terms. About the mid-eighteenth century, a Mr Roufe published the first table changing the words into numerical data and part of his table is reproduced on the next page.

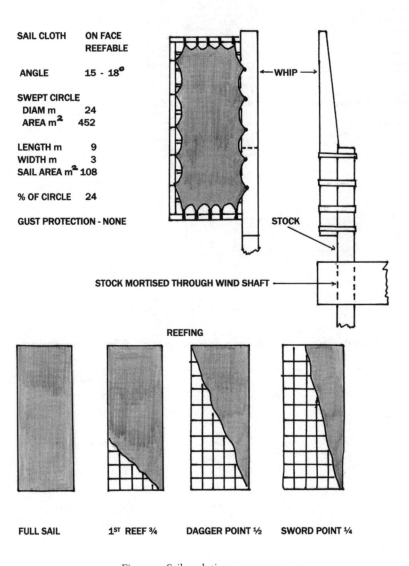

Figure 24. Sail evolution – common.

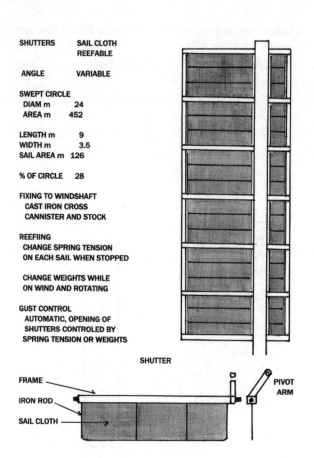

SHUTTERS　SAIL CLOTH
　　　　　REEFABLE

ANGLE　　VARIABLE

SWEPT CIRCLE
　DIAM m　24
　AREA m　452

LENGTH m　9
WIDTH m　3.5
SAIL AREA m　126

% OF CIRCLE　28

FIXING TO WINDSHAFT
　CAST IRON CROSS
　CANNISTER AND STOCK

REEFIING
　CHANGE SPRING TENSION
　ON EACH SAIL WHEN STOPPED

CHANGE WEIGHTS WHILE
ON WIND AND ROTATING

GUST CONTROL
　AUTOMATIC, OPENING OF
　SHUTTERS CONTROLED BY
　SPRING TENSION OR WEIGHTS

SHUTTER

FRAME

IRON ROD

SAIL CLOTH

PIVOT
ARM

Figure 25. Sail evolution –
spring and patent shutters.

ROUFE'S WIND SPEED

Speed mph	Force on 1 square foot in pounds	Wind Description
1	0	Hardly perceptible
5	0.1	Gentle wind
10	0.5	Pleasant brisk gale
15	1.1	
20	2.0	Very brisk gale
25	3.0	
30	4.4	High wind
35	6.0	
40	7.8	Very high wind
50	12.3	Tempest

As you can see, in words, a gale was less than a high wind, which is the opposite of today's terminology, but there is no doubt what the numbers mean in mph and pressure. The pressure data has been rounded, as the original figures were to three decimal places, indicating an unjustifiable accuracy. Later work also showed the higher speed data to be the least reliable, but it was a very commendable beginning and the force on the sails could be calculated. For a typical windmill sail of 26ft by 8ft (8m by 2.5m), the total force on all four sails if the sails were at right angles to the wind (area 832 sq ft) at 10mph is:

$$832 \times 0.5 = 416 \text{ pounds or } 0.2 \text{ tonnes}$$

The total force at other wind speeds is given below.

TOTAL FORCE ON SAILS AT VARIOUS WIND SPEEDS

Wind speed (mph)	Pressure (psf)	Full sail force (tonnes)	Sail	Resulting force (tonnes)
10	0.5	0.2	full	0.2
15	1.1	0.4	full	0.4
20	2.0	0.8	full	0.8
25	3.0	1.1	full	1.1
			¾	0.8
			½	0.6
30	4.4	1.7	¾	1.2
			½	0.9
35	6.0	2.3	½	1.2
40	7.8	3.0	shut down	
50	12.3	4.7	gust	

The sail could be reduced in area by reefing or rolling up the sail and tying it to the frame. This was the only option available to the miller for the common cloth sail and he could keep the resulting force on his sails reasonably constant, but he could only do this with the mill stopped. In sudden gusts, much higher forces were applied quickly to the sails, which he could not avoid, resulting in over speed which affected the fineness of the flour or, worse still, damage to sails, bearings gears and shafts.

What was needed was some system that reduced sail area as wind speed increased, to take less energy from the wind and keep the stones rotating as near to the optimum as possible. It had to be done while the mill was on wind and rotating, and be setable by the miller and keep that setting when the wind speed changed. The system must react fast enough to avoid over speed due to gusts and let the miller shut down the mill when the winds became too strong. The patent system had been established in 1617, making it financially attractive to exploit new ideas, and when explanations and diagrams were added to these they give an insight and understanding of the novel ways proposed to try to solve sail-speed control and alignment of the sails into the wind. Patent No.174 of 1674 by John Johnson proposed a new way of controlling the motion of a windmill, and in 1684 Nathan Heckford obtained patent No.243 for a horizontal windmill. Neither had details, but they show that the problems were being seriously studied. The disastrous effects of the 1703 gales, which destroyed 400 windmills, must have added urgency for the need to solve the problems of sail-speed control and alignment to the wind or luffing, but the solutions would be a long way ahead. In 1738 and 1744, patents were granted for horizontal windmills with eight sails which closed like doors to get drive from the wind on one half of the circle and opened to reduce drag on the other half. There was no gust control but the mill would operate from any wind direction and so need no luffing control.

In 1745, Edmund Lee of Brock Mill near Wigan in Lancashire, was granted patent No.615 for 'Self-Regulating Wind Machines'. The patent was valid in England, Wales, Berwick on Tweed and America. Lee is remembered for the invention of the fantail, but that was only one part of this notable patent. The other device was a weight-operated pitch control of the whole sail that rotated about a rod-like whip. The actual pitch was controlled by 'the regulating barr passing thro (sic) the centre of the original axis (of the wind shaft)' and set by weight, i.e. the larger the weight the greater the pitch and the greater the power extracted from the wind (Figure 28). When the wind gusted or strengthened, the increased pressure on the sail face partially overcame the pull of the weight and reduced the sail pitch, reducing the power taken from the wind and so tending to keep the rotational speed of the sails nearly constant. The concept was brilliant, but the engineering needed to build such units was still more than a century away. Lee had, however, provided the key to 'on wind and rotating control'. This was the idea of a rod sliding down a central hole in the wind shaft, to transfer mechanical movement of a lever, to change the extraction of power from the wind while the mill was operating. Lee owned and ran Brock Mill, so he may have experimented with variable pitch sails. There is, alas, no record of a fully operational mill and automatic reefing of the sails remained a dream.

In 1749 another patent was granted for a horizontal mill, to Richard Langworthy, whose system used eight vertical boards as sails but enclosed them in a ¾ cylinder whose ¼ opening was directed to face the wind by a tail vane. This avoided the need for the boards to rotate, so this would have been mechanically easier to construct and operate, but there are no records of one being built. While trying to solve the problems of the mill, the inventors had to guess at the forces involved and would not know how their new ideas would compare with existing mills, so it would be a very brave mill builder who decided to risk considerable capital in using these ideas. Possibly partly motivated by such needs, John Smeaton undertook an in-depth study of both water and windmills, which were the only sources of power at that time.

John Smeaton's study of windmill sails, presented to the Royal Society in 1759, not only dealt with the theoretical aspects of wind power but also recorded the standard design of sails in use at that time. In England, the sails were still rectangular and single sided with a weather angle for the whole sail of 15 to 18 degrees, whereas the Dutch already had a variable weather angle of 22 degrees at the poll end to 8 degrees at the tip with a divided arm. Such a sail was almost 40 per cent better than the simple sail so its adoption in England had begun. Smeaton did experiments with both and confirmed the superiority of the Dutch design and also experimented with a double-sided sail that increased in width from hub to tip – at the narrow hub end the leading edge was a board and at the wider tip end it was another cloth sail on a frame. While the tapered sail did not get adopted, it is possible that the short board became the full length weather board. The ratio of the widths of the double sails was between 1 to 3 and 1 to 4.5, with the narrow sail being on the leading edge. The angle on the wind board and the narrow sails can be greater than the angle of the main sail, so making a concave surface to the wind which Smeaton had also found to be beneficial in his tests. The common cloth sail was still the only type used, but that would now begin to change.

The first radical change in sail construction was by Andrew Meikle who, in 1772, invented the spring-operated shuttered sail (Figure 29). The shutters were thin boards or sail cloth stretched over a frame to which they were stitched and nailed. They were set at right angles to the sail stock, with three in a bay and six to eight bays on each sail. Each shutter had a short arm to close or open them, and these arms were linked to a bar that moved all the shutters on one sail at the same time. A spring held the shutters closed and the force on the spring could be adjusted. Springs were coil, full or half leaf, and it needed only 225mm of movement from fully open to fully closed, so either occurred very rapidly. On rainy days, rapid closure could be easily seen.

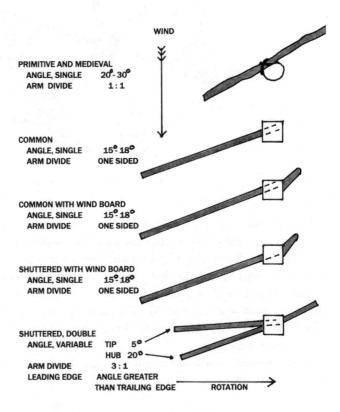

Figure 26. Sail profile.

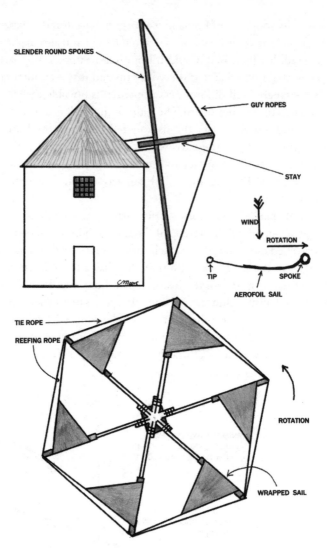

Figure 27. Mediterranean sails.

When the lower sail passed in front of the mill, the spring 'snapped' the shades closed throwing a shower of water onto the mill. The shades equally and rapidly opened when the sail cleared the mill. This phenomena was due to the fact that the mill structure interfered with the flow of the wind at this position, reducing the pressure on the sail and the power taken from the wind. It can affect the sail for 30 to 40 degrees of rotation out of 360, so reducing the power available by some 10 per cent. The closing of the spring sails in the lower position does compensate a little for this and is, hence, beneficial.

Meikle probably took out a patent in Scotland to protect his invention, but Scottish patents are unnumbered and hence extremely difficult to find. However, although he

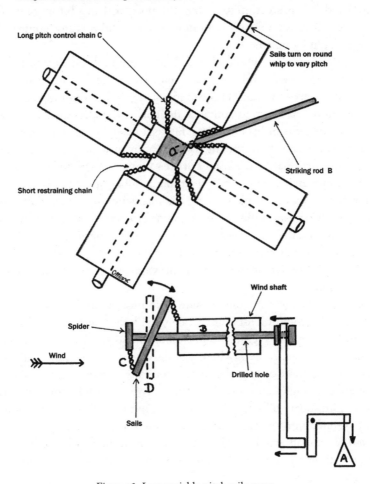

Weight A , via levers, pushes striking rod B up the hole in the wind shaft . Long chains C rotate the sails into the wind which sets them in motion . Gusts increase pressure on sails which overcomes the force of the weights and sails rotate out of the wind to position D , taking less energy from the wind so the sails should turn at a "constant " speed . The weights can be changed on-wind and the sail angle set manually .

Long pitch control chain C

Sails turn on round whip to vary pitch

Striking rod B

Short restraining chain

Wind shaft

Spider

Wind

Drilled hole

Sails

Figure 28. Lees variable pitch sails, 1745.

patented in England a machine for dressing and cleaning grain in 1768 (No.896) when he owned and operated Houstan Mill near Dunbar, and again patented a machine for separating corn from straw (No.1,645) in 1788, there is no record of an English patent for his spring sails. Although each sail had to be adjusted individually, it could be done more quickly than reefing cloth sails. Very importantly, in gusting winds the higher pressure of the gust opened the shutters against the spring force, spilled wind in seconds and avoided over speed of the sails. The springs re-closed the shutters when the gust had passed. The weight and cost of these sails was much greater than common sails, but the protection against gusts and the maintaining of a more constant speed meant the cost was worth paying. Common sails, having a smooth cloth face, were more efficient, and for the same area gave the stones more power. Some mills therefore converted only one pair to spring sails to retain high power but have some protection against gusts. Others converted all four sails where they considered gust protection to be more important than a bit more power. This was a very significant step forward but not 'on wind and rotating' control that was really being sought by millers.

In 1774, John Smeaton, who had built Eddystone Lighthouse and had already introduced cast-iron gears and done his detailed study of sails in 1759, designed a cast-iron cross to carry five sails. This was installed at a flint crushing mill which he built near his home in Leeds to convince the conventional, sceptical and reluctant industry of the benefits of his new ideas. Five, six or eight sails will start at a lower wind speed than four sails and rotate faster at a given wind speed, but the power obtained is proportional to the total sail area whose maximum is a quarter to a third of the swept circle area irrespective of the number of sails. This natural limit is imposed by the effects of turbulence, drag and interaction of the wind flow between sails. Interestingly, Smeaton did not seek patent protection for either the cast-iron cross or five sails, and these new concepts needed time to be accepted and used. Very few were used in the following fifty years, but in the growth period from 1820 about seventy mills with five, six, or eight patent sails were built, but this was out of a total of up to 15,000 mills.

The cast-iron cross was the only way to hold more than four sails on the wind shaft, but it became common for four sails along with the alternative poll end canister suitable only for four sails (Figure 30 and Colour 31 and 32). As usual, there were two exceptions (already mentioned) where six sails were fitted by mortise through the wind shaft or by a three-way poll end canister. The canister was two short square tubes cast at right angles and offset so that the two sail stocks could pass through them and a third square tube fitted over the end of the wind shaft. Stocks and wind shaft were wedged into the tubes. Either of these fittings made a very strong joint for sails and wind shaft avoiding the cutting of slots in the wind shaft, reducing rain penetration with its attendant rotting problem. They became the standard method of fitting sails to shaft.

In 1787, Thomas Mead, who described himself only as a carpenter of Sandwich, Kent, took out patent No.1,628 'Regulator for Wind and Other Mills'. This carpenter describes four ways of 'regulating all kinds of machinery where the first power is unequal'. Wind as a source of power is decidedly unequal and the carpenter proves himself to be a very capable engineer with these concepts of keeping process speed automatically constant from the widely varying speed of the wind. Mead is credited and remembered for the invention of the ball-operated centrifugal governor to control the separation of the stones, which will be described later in Grinding, but this is only one of the four ideas in this patent which also has a method for automatically reefing sails as wind speed increases. The concept, shown in Figure 31, is for one sail only, so the wind shaft would be festooned with ropes to control four sails with each rope retained on the shaft by hoops or staples. Each rope would be the length of the wind shaft, plus the whip and the width of the sail – a total of some 12m. The outward movement of the governor balls was probably no more than 0.5m and this would severely limit the amount of reefing possible on a 2-3m-wide sail. Magnifying the governor movement by levers could double sail reefing, but spring assistance would be needed to maintain the pull on the ropes; an ingenious idea but difficult not only to install but also to maintain. Furthermore, it was a 'consequential control', i.e. the stronger wind made the shaft rotate faster, which activated the governor to reef the sails, spill wind and slow the wind shaft down. Realistically, the

The spring pulls on the adjusting plate and lever which pushes the shade bar up and closes the shades . With 5 holes in the plate and 3 pins on the lever, there are 15 settings of spring tension to hold the shades closed . Gusts overcome the spring tension and the shades open to spill wind . When the gust has passed , the spring tension closes the shades and it needs only 9 inches or 225 mm of movement of the shade bar to go from fully closed to fully open where they can be locked by engaging the hook in the first hole of the plate for shut down of the mill . Each sail has to be set individually by stopping the mill for each sail .

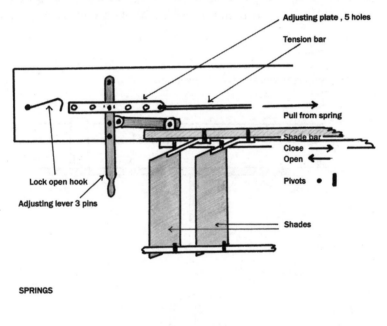

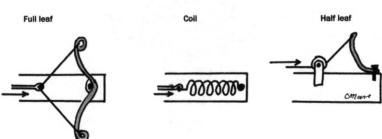

Figure 29. Meikle's spring sails, 1772.

control sequence would take a minute or so to complete and would be ineffective in dealing with gusts. Also, the pull needed to move wet sailcloth over a frame plus the friction on ropes at every turn, meant that the force from the governor would not be big enough for the job. It is therefore not surprising that there are no records of any of these systems being installed.

The adoption of spring sails by many millers must have encouraged Andrew Meikle to experiment further to get a system that controlled all the sails at the same time. In 1780, rumoured to be in collaboration with Smeaton – a powerful combination of knowledgeable engineer and practical millwright – the narrow shutters were changed to one rolled sail in each bay (Figure 32). They were rolled or unrolled simultaneously by a sliding frame, operated by centrifugal force to roll up the sail as the wind increased and a counter weight to unroll the sail as the wind slackened. The counter-weight system operated via a rod down the centre of the wind shaft. It would be a reasonable control in responding to varying winds but would be unable to rapidly spill wind in gusts. Friction on the sliding frame would be a serious problem, so apart from the trials on Meikle's mill no other units were

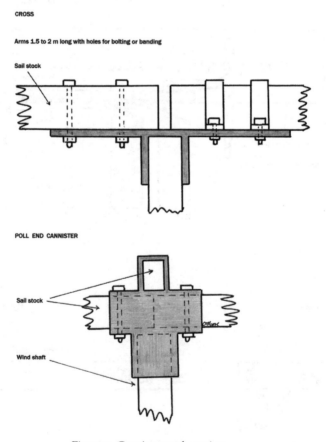

Figure 30. Cast-iron stock carriers.

REGULATION OF SHAFT SPEED IN VARYING WIND USING ROLLER SAILS

Sails are manually set by moving pulley Z . As wind speed increases the wind shaft rotates faster and drives the governor faster . Governor weights X moves outwards and raises collar A . This pulls rope B and collar C which pulls rope D and furls the longitudinal sails on their rollers so the wind shaft slows . The governor slows , governor weights X fall inward and collar A moves down . Return springs F pull collar C back so puling on ropes G which unfurls the sails from their rollers so more energy is taken from the wind so wind shaft and stones rotate at "constant" speed .

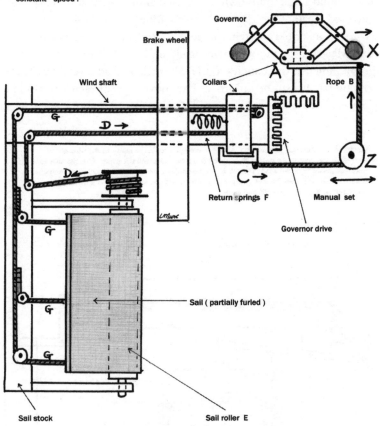

Figure 31. Mead's roller sails, 1787.

installed. It will remain a mystery why the centrally drilled hole down the wind shaft, striking rod, spider and cranks needed to operate the frames, were not applied to the shutters of Meikle's spring sails, which finally, some thirty years later, became the real answer to on-wind control with the ability to deal with gusts.

Catchpole invented an air brake for common sails. For the outer third of the sail, on the leading edge, a pair of longitudinal laths were hinged to two short side arms to form a wide wind board. In light winds, the laths remained closed (by springs) and so increased the sail area for more power. In gusts and stronger winds, the laths opened not only spilling wind but, by now being at right angles to the direction of travel of the sail, acted as a brake. It was automatic and effective and was used on a few mills in Suffolk. In the 1780s, Wiseman, Hilton and Heam got patents for a horizontal mill with a gust-control system, roller reefing and variable pitch control, but they were unable to persuade anyone to build a mill to their designs.

Captain Stephen Hooper of the Isle of Thanet was not only an inventor but also had the capital to build a mill to his own design to demonstrate that his ideas could work. He was granted a patent for 'on wind and rotating' sail control in 1789 (Figure 33). Patent No. 1,706 describes a system for cloth control addressing some of the problems involved in Mead's ideas. Hooper used small sails like Meikle's shutters, so instead of one large sail pulled at right angles to the sail stock, he had several (say fifteen) small sails operating parallel to the sail stock. The movement of each sail would be small and ropes were used only to pull the sails from rollers, i.e. about 1.0m. Mechanical

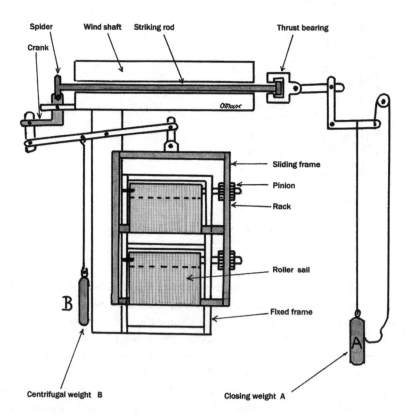

Weight A, via the rope, striking rod, spider and crank unrolls the sails. As the wind speed increases, the centrifugal weight B is pulled down the whip and via levers, furls the sails to take less power from the wind. Weight A can be changed by the miller to control the amount of sail rolled out. If the wind eases, weight A via the ropes and levers, pulls out more sail to maintain the speed of the sails and stones.

Figure 32. Meikle–Smeaton sliding frame sails, 1780.

linkage via several racks and pinions transferred the movement to a rod operating down the centre of the wind shaft, but only for a limited distance (say halfway). A slot in the shaft permitted the control rod to move a ring backwards and forwards along the outside of the shaft. The ring engaged a collar driving a rack and pinion which could be set by hand and held in position by weights. Control would be restricted due to the 'slack' in so many linkages that had to be taken up before any movement of the sails occurred, and the force needed to overcome the friction in so many pivots, racks, pinions and rollers would have been large. As well as the hand wheel there was a conical pulley on the same shaft and a weighted rope caused the sails to

REGULATING THE POWER OF THE WIND WITH ROLLER SAILS

Sails are set full by rope A and weight B holds sails unfurled. When wind strengthens sails turn faster and weight C moves outward by centrifugal force. Via rack and pinion D sails furl and via rack and pinion E the spider moves outwards, pulls the striking rod and moves ring F up the wind shaft. Linkage G operates rack and pinion H and moves weight B outward on the conical pulley so increasing the return force. When the wind eases, weight B operates the linkage and moves the sails back to fully unfurled automatically changing sail area to suit the wind strength.

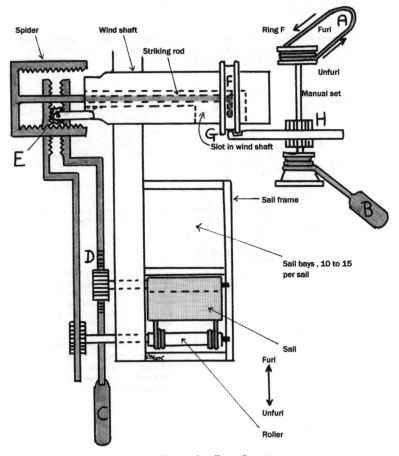

Figure 33. Hooper's roller sails, 1789.

unfurl to get maximum wind. As the wind speed increased and the sails turned faster, centrifugal force on weights at the end of the stock caused the small roller sails to roll up, reduce sail area and slow down the mill. A few were installed in Sussex, East Yorkshire and the Fylde area of Lancashire. However, Hooper's mechanism could not deal with sudden gusts, and the complicated linkages would need a lot of care to keep them adjusted and working properly.

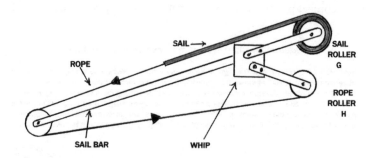

Gear wheels A and B will drive the rollers to furl and unfurl the sails. They are held on the wind shaft by collars and are normally stationary while the wind shaft rotates within them. There is a stud L on each. Lever J pivots on a projection from the wind shaft. Springs and centrifugal force

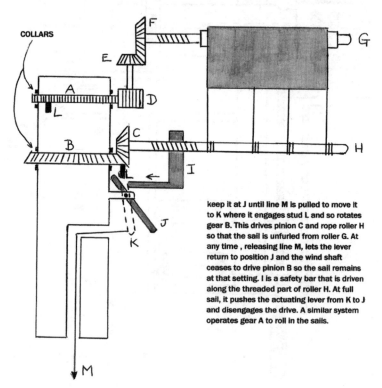

keep it at J until line M is pulled to move it to K where it engages stud L and so rotates gear B. This drives pinion C and rope roller H so that the sail is unfurled from roller G. At any time , releasing line M, lets the lever return to position J and the wind shaft ceases to drive pinion B so the sail remains at that setting. I is a safety bar that is driven along the threaded part of roller H. At full sail, it pushes the actuating lever from K to J and disengages the drive. A similar system operates gear A to roll in the sails.

Figure 34. Bywater's roller sails, 1804

Hooper was also involved in the building of two horizontal windmills and George Silvester was granted a patent for a horizontal mill where the eight sail panels rotated on pivots on a square chain that itself turned to drive the mill. It was this system that may have powered a clay-grinding mill on the Duke of Bedford's estate but these novel windmills were very unusual, and, as Telford's survey of 1796–98 showed, the overwhelming majority of mills being built were still using four common sails without a fantail to bring them automatically into the wind.

Bywater gained a patent for another roller-reefing system in 1804, where the power of the wind was ingeniously used to reef and unreef the sail (Figure 34). A company was set up to exploit the patent, but the system offered no gust control and only a few were installed in Lancashire. The milling industry remained sceptical and was not convinced that horizontal mills and roller reefing were acceptable solutions – the standard was still the four-common-sailed mill with a few others fitted with two common and two of Meikle's spring sails and some with four spring sails.

Then, in 1807, William Cubitt of Norfolk perfected his system of automatic control and took out patent No.3,041 (Figure 35). It changed Meikle-type shutters from spring to weight control to hold them closed. The striking rod passed completely down the central hole in the wind shaft and operated toothed racks and pinions to pull the shutter bars and close the shutters (Colour 33). The system was much simpler than Mead's, Hooper's or Bywater's, so less force was lost to friction and less slack would need to be taken up before the shutters opened or closed. Whilst the patent describes racks and pinions to operate the system, the wording of the patent is astute so as not to exclude other methods of linkage. Cubitt came from a family of millers in Norfolk, but local rumour has it that his very traditional father would not have his son's new type of sails on his mill. So Cubitt moved to Stalham where Cooke's smock mill was the first to be fitted with his sails and Cubitt married Cooke's niece about two years later. The first post mill to use Cubitt's sails was at Horning, and by 1815 about twenty mills in Norfolk had installed them. This phenomenal rate of uptake shows how good the system was, and at an early stages the rack and pinion must have been changed to the weighted lever fixed to the fantail supports (Figure 36).

The lever version is simple engineering with the minimum of moving parts. Compare this with the mechanisms of Mead and Hooper in Figures 31 and 33. It was therefore reliable in working, easily set or altered to suit the wind and, critically, it could spill wind during gusts and avoid over speed. From stationary with shutters open, it was an easy ground or balcony task to pull the chain to close the shutters and set the mill going. Experience told the miller how much weight to fix to the chain to hold the shutters in the wind that was blowing. There was a nest of three cast-iron weights shaped like square buckets (Colour 34). The largest was 200mm^2 at the top, tapering to 75mm^2 at the bottom, weighing 20kg, and it is usual to have this bucket on as a minimum. The middle weight fitted into the larger bucket and weighed 10kg and the smallest could be hollow at 5kg or solid at 10kg, fitting into the middle bucket so giving a weight range of 20 to 40kg on the chain. The chain was fixed to the striking lever which is 2 to 2.25m long and protrudes from the rear of the cap in line with the wind shaft.

At the inner end of the striking lever, a stub, at right angles to the main lever about 225mm long, is fixed to the back of the striking rod which passes up the centre of wind shaft. The 1.25 to 1.5m movement of the end of the lever (which is also the movement of the weight bucket) is reduced to about 230mm for the end of the stub and the striking rod – a ratio of about 6 to 1. The force from the weights is multiplied by the same ratio so the 20 to 40kg weights become a force of 120 to 240kg (0.12 to 0.24 tonnes).

At the sail end of the striking rod is the spider which is linked by four cranks to the shutter bars on each sail (Colour 33). The two arms of the cranks are equal so the 230mm movement of the striking rod becomes a 230mm movement of the shutter bars which, in turn, are fitted to the 150mm-long arms on each shutter. It requires about 90 degrees of rotation of the shutter arm to take a shutter from fully closed to fully open. This is a ¼ of the circumference of a 150mm-radius circle which is about 230mm.

For the typical windmill, with four sails each 8m by 2.5m, i.e. 80m² of sail, the full force on this area at right angles to a 25mph wind is 1.1 tonnes, whereas the maximum holding force from the weights is only 0.24 tonnes, yet this force will hold the shutters closed! This apparent anomaly is explained by the aerodynamics of the winds interaction with angled rotating sails. For a start, the sails are not at right angles to the wind and, as shown earlier by reference to ships sails, a sail at right angles feels the full force of the wind but one parallel to it feels none. The windmill sail is 15 to 18 degrees out of the wind, which reduces the force on the sail but still causes rotation which takes energy from the wind and reduces the force even more. In the early days of the patent sail, various weights and leverages would have been tried to find the practical answer of 6 to 1 leverage and 20 to 40kg of weights which is still used today to hold the shutters closed up to wind speeds of 15 to 25mph. Above these speeds, the shutters opens to a degree dependent on the higher wind speed to let wind pass through the sail rather than taking power from it to keep the sails turning at the speed the miller wants. Gusts rapidly open the shutters fully and avoid increase in speed and its consequential damage to the mill. The weights equally rapidly close the shutters after the gust has passed and restore normal working. By changing the weights, the miller can keep his mill going to suit the wind speed, number of stones in use and the rate of grain feed. Shut down was equally simple. Remove the weight, pull on the opposite chain to fully open the shutters spilling all the wind and bringing the mill to a halt. The solution to dealing with gusts and 'on wind and automatic' control had been achieved.

Cubitt moved on in engineering, getting involved in the construction of canals, docks, railways and finally, in 1851, the building of the prestigious Crystal Palace for the Great Exhibition for which he was knighted. His sail patent expired in 1821 and the system became known simply as 'patent sails', and perhaps this is fitting as it recognises all the engineers who contributed novel ideas over some sixty years in the attempts to solve the problem of on wind and rotating control with instant wind spill in the event of gusts.

Weight A pulls rope and turns pinion B which pushes striking rod C up the wind shaft. Spider
D moves out and pushes rack E in. This rotates pinion F and arm G which pulls the shutter
bar H and shutters J close and the sails turn. In gusts, the wind pressure opens the shutters and
spills wind. When the gust has passed, the force of weight A closes the shutters. Weight A can
be changed " on wind and rotating " to fix the force that closes the sails so taking reasonably
constant power from the wind to keep the stone speed steady.

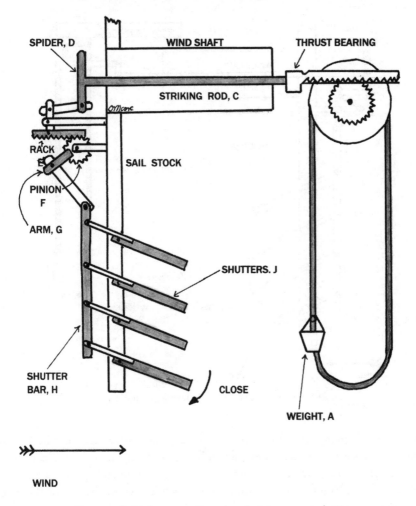

Figure 35. Cubitt's patent sails, rack and pinion operated, 1807.

In the 1820s, the rapid expansion of windmill building began to meet the
increased demand for flour for a rising population. The four-sail mill remained the
norm but five, six and eight-sailed mills were built, yet out of the thousands of mills
constructed, only sixty or seventy were multi-sailed mills, and of these less than
ten were eight sailed. The common sail was, by now, rarely installed on new mills
and there was, perhaps, a natural progression for existing mills. When common sails
needed replacing two were retained and two spring sails added. The next change

Weight A pulls rope and lever B down. This pushes the striking rod C up the wind shaft. The spider D moves out and via the crank E pulls the shutter bar F and closes the shutters J. In gusts or stronger wind, the wind pressure opens the shutters and spills wind. When the wind eases, weight A , via the linkage, closes the shutters again. Weight A can be changed to suit wind strength and so control the speed of rotation. Sails can be stopped by pulling the return rope H which opens the shutters fully spilling the wind and the sails cease to turn

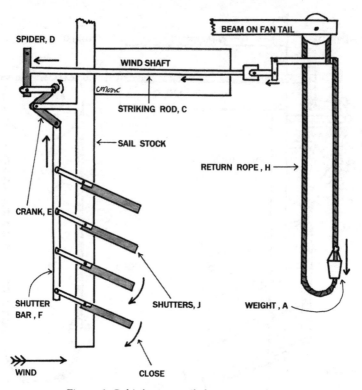

Figure 36. Cubitt's patent sails, lever operated, 1807

could be to all spring, and finally to all patent. New mills employed the best sail design, which was the twisted patent sail with an arm divide of 3 to 1, but the change to this was not countrywide. The south preferred common and spring or all spring, while the west retained the most common only sails. Down the east coast, the patent sails were the norm.

One other innovation was introduced, probably early in the nineteenth century since it is not mentioned in Telford's study of 1796, and that is the slight inward coning of the sails. The face of the whip in contact with the cross or stock is cut at a small angle, say 5 to 8 degrees to give a cone angle of 10 to 16 degrees which improves the efficiency of the sails. Its adoption is difficult to quantify since it is not easy to see such small angles, and only the detailed millwright's records would reveal what the angle was on a particular mill.

By 1841, the behaviour of the wind had been further quantified and the following accurate data published by Sir Richard Phillips:

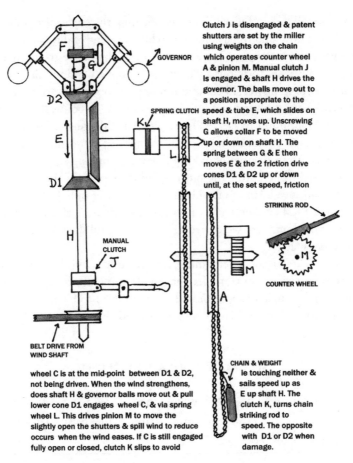

Figure 37. Hammond patent control.

At 12mph wind speed, the wind pressure is 0.75 pounds per square foot

At 25mph wind speed, the wind pressure is 3.125 psf

At 50mph wind speed, the wind pressure is 12.5 psf

The wind pressure is quadrupling as the wind speed doubles, and for the usual mill with four sails each 8m long and 2.5m wide, these apparently small numbers mean that the total nominal load on the sails at 12mph is about 0.3 tonnes. A 25mph gust increases this to 1.2 tonnes and a 50mph gust generates a thrust of 5.0 tonnes! (See Appendix III for the calculations.) These forces are big and it is now very obvious why gusts cause the mill to over speed and result in serious damage unless the sail control rapidly spills wind. Both spring and patent sails can do this, and patent sails can be reefed on wind, so it is obvious despite the regional variation why these sails became so widely used, particularly in the boom period of mill building from 1820 to 1870. Furthermore, after Cubitt's patent of 1807 there were no more patents or inventions on sail design for windmills, confirming the superiority of the patent sail to spill wind in gusts and to be adjustable while on wind and rotating.

There were, however, one or two novel ideas that were tried. In 1868, Henry Chopping and Warner took out a patent for a mobile windmill that could be taken to any site needing power, like a traction engine, but there are no records of any being built. A very sophisticated fine tuning system for controlling patent sails was patented by Hammond in 1873 (Figure 37). An additional governor adjusted the position of the patent shutters in small changes of wind speed that the weights would not react to in an attempt to keep the stones rotating at a truly constant speed. It was supposed to be automatic but needed to be reset if there was a significant change in wind speed. He installed it in his own tower mill, Jack at Clayton in East Sussex, and at the Herstmonceaux post mill in which he also had an interest. It is not known how long they were used or how successful the system was at keeping a constant stone speed.

In the 600 years of sail development, the area of the sails had increased from less than 10 to over 100m², sail fabric had changed from cloth to shutters and the angle to the wind from single to sophisticated twist. All this had been done to increase the milling capacity and these changes are summarised below:

SUMMARY OF SAIL EVOLUTION

Type	Century	Length (m)	Width (m)	Area (m²)	Angle (degrees)	Arm Divide	Stones (No.and diameter)
Primitive	12th/14th	2	1	8	15	1 to 1	1; 0.6
Medieval	13th/16th	4	1.5	24	15	1 to 1	1; 1.2
Common on stock	14th/17th	5	2	40	15	1 sided	1, 2; 1.2–1.5
Common on whip	15th to present	9	3	108	15	1 sided	3,4; 1.2–1.5
Spring shutters	1772 to present	9	3	108	5 to 20	3 to 1	3,4; 1.2–1.5
Patent shutters	1807 to present	9	3	108	5 to 20	3 to1	3,4; 1.2–1.5

The data in the preceding table should be taken as a reasonable guide to when and what changes occurred to the sails, but there are variations as there are to all aspects of windmill figures, and further details on the diversity of the number and diameter of the stones are given in the section on milling.

Basic data for the most used sails is summarised below:

Number:	4
Rotation:	anticlockwise
Design:	Patent, double–sided
Arm divide:	3 to 1
Length:	9m, 3 to 4 times the width
	85 per cent of arm length
Width:	3m
Sail area:	108m², 25 to 33 per cent of the swept circle
Bays:	6 to 8
Shutters:	3 per bay, canvas on frame
Angle:	20 degrees at hub
	5 degrees at tip
Weight:	750kg to 1 tonne each
	3 to 4 tonnes total
Coning:	10 to 16 degrees

Performance:

Wind speed range:	8 to 40mph
Preferred speed:	15rpm (12 to 18rpm in 25mph wind)
Power:	12 to 16hp for up to four pairs of stones
Efficiency:	about 20 per cent
Tip Speed Ratio*:	2 (1 to 3)

* Tip Speed Ratio (TSR). The ratio of the wind speed and sail tip speed is regarded as the modern way of expressing sail efficiency. However, in 1796, Telford discussed using these relative speeds to get the best performance but did not call them TSR. See Appendix VI.

CHAPTER 9

The Power Train

The simplest power train, as one would expect, followed from the direct drive of the hand quern where man's (or more usually woman's) arm provided the power via a simple peg in the top stone. The Seistan vertical windmills were also directly driven from the sail shaft, as were the stones in horizontal waterwheel mills and Roman beast mills where the ass, cow or horse walked in a circle harnessed to a beam that drove the mill stone at 3 or 4rpm. Gearing became necessary firstly in watermills, when the wheel became vertical and drove a horizontal shaft. A vertical shaft was required to turn the horizontal mill stones. This transition was accomplished by fixing a large diameter wheel on the horizontal shaft, on which wooden pegs engaged staves on a lantern pinion to drive one pair of stones. The same principal was used on the early post mills (Figure 38).

The horizontal wind shaft soon became inclined at 8 to 15 degrees, was made of oak some 375mm² with four sails mortised and wedged into it. Just behind the sails, a groove 25mm deep was cut into the shaft to trap rainwater and prevent it running down the shaft into the mill. Inside the cap is the front or neck bearing, 225mm diameter, made from oak, sometimes lined with stone – particularly basalt or marble. This bearing (Colour 35) takes the main weight of the shaft and sails and two or three wrought-iron bands were let into this part of the shaft to reduce wear. The shaft then tapers towards the rear end where the tail bearing is strapped to the tail beam to absorb the thrust on the shaft generated by the force of the wind on the sails (Colour 36). To get more production a second, slightly smaller wheel, was added in the Middle Ages at the rear end of the shaft to drive another pair of stones. The older wooden wheels were made of elm sections for a stout rim into which holes were made by augers for the drive pegs. Oak spokes, called compass arms, then joined the rim to the shaft by mortising so the wheel was fixed in one position on the shaft, but the lantern pinion was very tolerant to lack of alignment of shafts and depth of engagement of pegs, so the inability to move the wheel on the shaft was not a problem.

The big front wheel became known as the brake wheel after the brake was introduced in the early 1600s, and the smaller rear wheel was called the tail wheel. The introduction of cogs, cants or teeth on the wheels and wallower meant that there had to be a much more accurate and adjustable way to get the correct meshing. The original way of spoke fixing does not permit any lateral adjustment of the wheel. The answer was the 'clasp arm', where the wheels were made separately from the shaft and slid along it from the rear end and adjusted to mesh properly and then wedged (Figure 39). These wheels are 2.1 to 3m diameter with arms 250 to 375mm² supporting a rim 225mm wide. This type of construction continued even after the introduction of cast iron. Cast-iron brake wheels are similarly wedged onto cast-iron shafts.

The wallowers in the post mill rotate the stone drive shaft which moves on a pivoted arm so that the wallower can be put in or out of gear. This can only be done when the sails are stopped – trying to engage the stationary wallower into the rotating brake wheel would cause major damage and probably strip the teeth from the wheel. Apple, hornbeam or beech was used for the teeth, which had a pitch of 125mm. The brake and tail wheels had sixty to 100 teeth and the wallowers twelve to twenty, so the increase of speed of the stone shaft was about five times the speed of the wind shaft. From a small number of post mills where specific data was available, the speed-up ratio ranged from 4.8 to 5.4. In a 20mph wind with sails turning at 15rpm the stones would be spinning at about 75rpm – an improvement on the best speed for hand querns of about 60rpm.

Once cast-iron gears began to be used in the 1750s, the wooden cogs could be replaced by this new stronger material which could transmit a greater force. The pitch was reduced to 75mm so more cogs could be used on a given diameter. Cast-iron cogs could be added to a wooden brake wheel by bolting six to eight segments of cogs to the face. A 2.4m-diameter brake wheel would have sixty wooden teeth at 125mm pitch, which could be increased to 100 cast-iron teeth at 75mm pitch. The wallower could be increased in diameter to improve its strength but wooden teeth were retained for quiet and lubrication-free meshing of the cogs but with more cogs on the brake wheel and the wallower speed up ratio remained essentially the same as for all wooden cogs.

The higher tower and smock mills with more floors and without the inconvenience of a large post in the centre could use a different and more versatile power train so that more stones could be driven (Figure 40). In these mills, the brake wheel drives the permanently meshed wallower (Colour 37), rotating on a shaft at the centre of the mill which passes down through the dust and bin floors to the stone floor. On this shaft, in the roof of the stone floor, is the great spur wheel (Colour 38), into which three or four pinions (called stone nuts) can be put in gear to drive three to four pairs of stones. Some had enough space for six pairs of stones but three was more usual. This common form of drive is called overdrift (Figure 40). The alternative is underdrift (Figure 41), where the central shaft passes through the stone floor and the great spur wheel is near the ceiling of the meal floor. The spindles from the stone nuts go back up into the stone floor and drive the stones from below. In the nineteenth century, this set-up was used in a few post mills so that two pairs of stones could be driven from the brake wheel with the tail wheel directly driving the third pair.

Even when cast iron became available, it did not entirely replace the traditional oak wind shafts and wheels. An oak shaft 600mm² at the front tapering to 300 to 375mm at the tail, could be changed to a cast-iron shaft of 300mm diameter at the sail end tapering to 150mm diameter at the tail. These shafts could be 2.4 to 4.5m long and weigh 1 to 5 tonnes with a central hole of 50 to 75mm diameter to take the striking rod for the control of patent sails once they were introduced in 1807. The front bearing area of the iron shaft becomes 250mm diameter, 225mm long with 50mm-thick brass or gun metal sleeves as the liner in the bearing housing which generally remained a half or cup design so that the weight of the shaft and sails hold the rotating shaft in position. The tail bearing becomes a substantial iron pot strapped or bolted to the tail beam and incorporates iron or brass pads to take the large thrust caused by the wind on the sails. These developments significantly increased the life and smooth running of the mill.

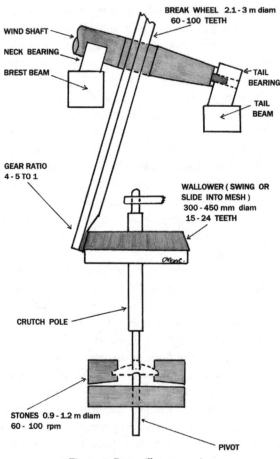

Figure 38. Post mill power train.

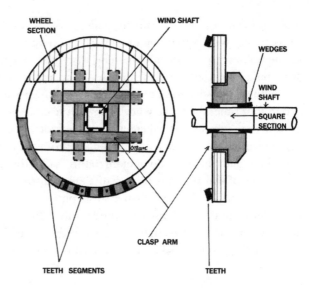

Figure 39. Clasp arm brake wheel.

An iron brake wheel would be 2.1 to 3m diameter, 100mm wide with six to eight spokes and integral cogs. The alternative was a duplex structure of an oak clasp arm wheel 225mm thick with iron cogs cast in segments that were bolted to the rim. In both cases, the pitch of the cast-iron cogs would be 75mm and the range of cogs on iron brake wheels is 60 to 120. The centre of the wheel has a square hole which fits over the square section of the wind shaft. The wheel can be pushed up from the tail end, where the diameter is smaller, onto the square section which is 12 to 25mm less per side than the hole in the wheel. Oak wedges are hammered in from both sides to hold the brake wheel in the right position to mesh with the wallower on its vertical shaft. The inclination of the wind shaft at 8 to 15 degrees means that the cogs on the brake wheel and wallower are also angled to accommodate this tilt so the wallower has become a bevel or crown wheel.

The brake wheels are big and the force they transmit large, so the braking system (Figure 42) has to be substantial. Four to six shoes of ash or elm, 100mm thick and the full width of the rim, are linked by iron arms and pivots and almost completely encircle the brake wheel. The brake lever is a 3m-long, 150mm² oak beam or an iron girder of the same weight, which will close the shoes onto the rim to slow the wheel and the direction of rotation of the wheel tends to increase the braking force. The heavy brake lever (Colour 39) is normally held up and the brake off by a hook engaging a rod in the end of the arm. To put the brake on, the miller jerks the brake rope which disengages the hook and the weight of the arm applies the brake. If the mill is tail winded by a passing storm (or an inattentive miller) the sails can rotate in the opposite direction. With this type of brake, the force from the shoes is reduced by such rotation, so it is very difficult to stop the mill and considerable damage could result. As iron became available, the wooden shoes could be replaced by an iron band which tightened round the rim for braking. In both cases, the rubbing action of the shoes or band polishes the rim and reduces the braking force. Sand or grit thrown between the two roughens up the surface and reinstates braking.

The wallower, permanently meshing with the brake wheel, is half to a third of its diameter and can still be made of oak or elm, but is more likely to be made of cast iron with twenty-five to sixty cogs, so the resulting speed increase of the wallower is two to four times the brake wheel. The shaft from the wallower goes centrally down the mill to the stone floor where its lower end rotates in a bearing block fixed to one of the main beams or into a post coming up from the foundations of the mill.

In the roof of the stone floor, fitted to this shaft is the spur wheel made of elm, iron or a duplex of elm and iron. It is 1.7 to 2.8m in diameter with 65 to 140 cogs on a 55 to 95mm pitch. Two, three, four, five or six pinions (called stone nuts) can engage the spur wheel, but the more usual number is three. The stone nuts are 375 to 600mm in diameter with 15 to 27 teeth so there is a further speed increase of four to five times. The overall speed increase from wind shaft to stones deduced from some forty mills

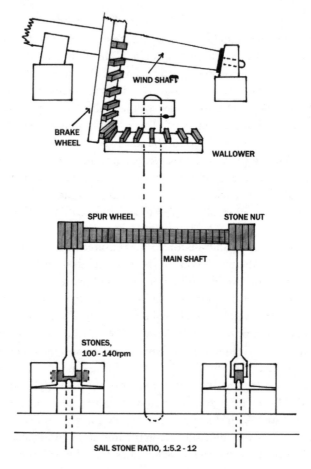

Figure 40. Tower and smock mill power train – overdrift.

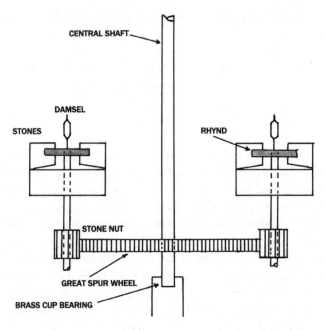

Figure 41. Tower and smock mill power train – underdrift.

is 5.5 to 12. For the majority with four sails the ratio is 5.5 to 11 with an average of 8.3. As the number of sails is increased to five, six or eight there is a natural tendency for such sails to rotate more slowly than four sails, so a higher ratio is needed to keep the stones rotating at their desired speed. There never were many such mills, but from very limited data the change in gearing can be illustrated.

No. of sails	Average stone to sail speed ratio
4	8.3
5 & 6	10.3
8	11.4

Stone diameter has also to be considered when deciding the gearing ratio. At the same speed of rotation a larger diameter stone has a higher peripheral speed, which is beneficial in making finer flour but to maintain the same peripheral speed a lower ratio would be needed. From a comparison of data from a reasonable number of mills, the overlap in the range of ratios indicates that it was not of major importance for 4 to 5ft-diameter stones, as the table on the following page shows.

Stone diameter		Stone to sail speed ratio	
ft	m	Range	Average
4	1.2	4.8–12	8.4
4.25	1.3	8.0–9.0	8.5
4.5	1.4	4.8–10.8	8.7
5	1.5	6.6–8.4	7.5

For the overwhelming majority of mills with four sails and three stones of 4ft diameter, it is reasonable to conclude that the typical stone to sail speed ratio was 8.4 which could be achieved with a very large combination of cog numbers on brake wheel, wallower, spur wheel and stone nut to drive such stones at about 120rpm. Some mills, however, do have stone nuts of different diameters with different numbers of teeth, so that larger stones are driven more slowly than their smaller cousins to get the same peripheral speed. In other cases, for mill stones of the same diameter different-sized stone nuts were used to give higher speeds to particular stones. The increase in speed was small at 10 to 15 per cent, but such an increase would give finer flour. It is also recorded that burr stones for flour should be driven faster than Peak stones for feed, but the contrary view is also found. The logical conclusion is that good flour can be produced over a range of speeds and this aspect will be considered in the chapter on milling.

The stone nuts themselves are set on an oak or iron shaft which can be slid, lifted or swung in or out of mesh with the spur wheel, which must only be done when the sails are at rest. The miller has to decide how many – and which – stones he needs before milling begins, and at the end of milling he will run them clear of flour, disengage the stone nuts and separate them. He may, however, decide to leave one set of stones in gear so that if strong winds overcome the braking system he can feed grain into this set of stones which will act as a large disc brake to slow the mill down and avoid serious damage.

The number of cogs on meshing wheels is well thought out. If the number of cogs on one wheel divides equally into the number of cogs on the other, e.g. spur wheel 125 teeth and stone nut twenty-five teeth, the same cogs will mesh five times in each revolution. If one cog cracks, chips or wears, it will rapidly damage the corresponding five cogs that are engaged each turn. If the spur has 125 cogs and the stone nut twenty-four or twenty-six, it is several hundred revolutions before the same cogs mesh and wear is evened out. The additional or omitted cog is called the hunting cog. Wood-to-wood teeth run quietly and without lubrication, whereas iron-to-iron teeth need permanent lubrication. It was soon learned that wood on iron teeth mesh and run more quietly than iron on iron and wooden teeth soon bed into the iron ones and will wear preferentially, preserving the more expensive iron teeth on segments and, more importantly, teeth cast into the wheel itself. It is therefore usual to have wooden teeth on the stone nuts since, when they do wear, it is an easy and low-cost task to knock out the worn wooden cogs and replace them.

Tower and smock mills had either an over or underdrift drive depending on the regional practice of the millwright. There is no particular advantage to either, but overdrift seems to have been the most popular. Post mills were traditionally overdrift and only rarely did they have spur wheel and stone nuts. The final drive for overdrift stones is by the stone shaft, crutch pole or quant (Figure 43), which was originally oak but became iron of 60 to 100mm² section. The lower end of this is shaped into two prongs that slide over the rhynd and engage the mace, which is fixed on the square tapered section of the spindle just below the domed nose. Slots in the mace engage the rhynd and two other slots take the prongs so the spindle rotates with the upper stone. There is slack between the prongs and the mace to let the prongs and its shaft move a little to let the stone nuts be moved in or out of gear with the great spur wheel. For underdrift drive, a square section on the spindle engages a square hole in the rhynd (Colour 40).

The mills were built to operate in wind speeds of 8 to 40mph with sails rotating at 12 to 20rpm. As wind speed increases, common sails were reefed to reduce sail area and take less power from the wind. With spring sails, the spring tension was reduced to let the shutters partially open in a strengthening wind, spill some wind and again keep the sails rotating at the best speed. Both required that the mill be stopped for every sail, so the best control was achieved with patent sails where the weights that hold the shutters closed can be changed at any time without stopping the mill. A further option is to select the number of stones and rate of feed in use, i.e. in light winds use one or two pairs and in stronger winds use three or four pairs. The mill has, of course, to be stopped to engage more stones, and even with four pairs at work at wind speeds of 25 to 40mph sail speed may exceed 20rpm, and at 40mph wind the mill must be stopped to avoid damage to sails shafts and drives.

Milling will therefore take place between 8 and 40mph wind (with appropriate reefing) to get sail speeds between 12 and 20rpm and stone speeds are in the range of 90 to 140rpm. The usual speed for French burr stones grinding flour and Derbyshire Peak stones milling feed is about 120rpm. Typically in a 20mph wind with the sails turning at 15rpm the mill is working at 10 to 15hp (about 12kW) to rotate three runner stones each weighing 1 ton. In gusts the power delivered by the sails can rapidly double to 30hp, so the power train has therefore to be very robust.

The power train can also be used for some very important secondary operations. Since stone speed cannot be held constant, milling is controlled by regulating the gap between the stones as the speed varies. This is done by the tentering gear which is operated by a governor driven from the central shaft or stone spindle at about 60rpm. The operation of this vital piece of milling equipment is dealt with fully in the chapter on Milling. All smock and tower mills had another useful driven device and that was the internal sack hoist (Figure 44). Grain sacks weighed 280lb (135kg) and the stairs of the mill were narrow and steep and the sacks had to taken to the top floor for the milling process to begin. The sack hoist was an endless chain (sometimes a rope) driven by friction from the wallower. A roller or a 'Y' wheel (Colour 41) carried the chain that passed down through all floors to the ground. The sacks were hooked to the chain which pulled them through upward-opening double trapdoors on each floor. The trapdoors fell closed after each sack had gone through, ensuring that if a sack fell off it could only fall one floor. This design also ensured that the trapdoors were normally closed so the miller could not fall through either. A few post mills also had sack hoists driven similarly by friction from the wallower, but these were external, acting more like a crane, with the sack swung into the door and dropped on the floor.

Grain cleaners which turned at 20rpm were usually driven from the central shaft and located on the dust floor. On the meal floor, flour dressers or bolters were used to remove the bran and grade the flour – this needed speeds of 50 to 60rpm or 300 to 400rpm depending on the type, and the higher speed units could produce several grades of finer whiter flour that became increasingly popular from the 1850s onwards. Fans needing similar speeds were also introduced to provide suction to get more rapid removal of flour from between the stones for higher production, but these were rare in windmills. All these machines are dealt with in more detail in the section on milling, to which their proper function was vital for the production of good and consistent quality flour.

The major drawback to consistent and regular operation was the variability of the wind. It only blows in the usable speed range of 8 to 40mph for about a quarter of the time, with long and unpredictable periods of flat calm and shorter gales, neither of which allow milling. When steam engines became available from the early 1800s

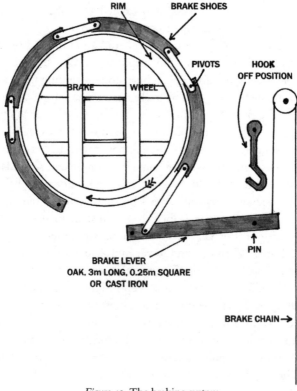

Figure 42. The braking system.

and were used in steam-powered-only mills, the wind millers began to look at this novel power to drive the stones in their tower mills. A minority of mills installed engines of about 10hp to drive two pairs of stones. Boiler and engine were housed in a shed close to – but separated – from the mill, so that the boiler fire could not endanger the mill. The power from the engine was transferred to the mill by a boxed ground-level shaft with a pulley and 150mm-wide leather belt that drove a flat pulley wheel on the outside of the mill at meal or stone floor level. In other cases and with mobile engines, the belt went from the engine directly to the high-level pulley. A shaft went through the mill wall where a train of cast-iron gears took the power to the spur-wheel shaft. The final drive of the central shaft was a gear wheel dedicated to this power input. The spur wheel remained as the drive for the stone nuts, though sometimes the steam-drive gear wheel was bolted to it.

When the wind had dropped and the steam engine was to be used, a removable section of 5 to 7 cogs was taken out of the brake wheel to let the wallower rotate with the central shaft without turning the sails. Steam power let the miller grind corn on a regular and consistent basis. He could mill 8 to 10 hours in daytime, five or six days per week, to keep his customers supplied. Even with only two pairs of stones driven by steam, this gave 80 to 120 hours of production time per week.

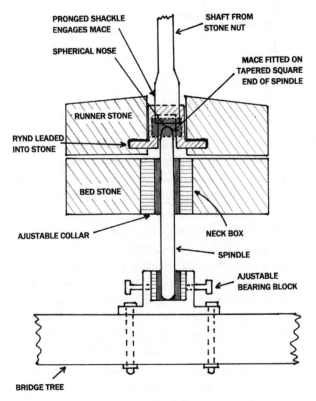

Figure 43. Final drive.

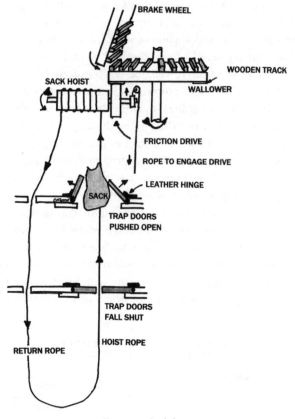

Figure 44. Sack hoist.

In wind, with three pairs operating on average only a quarter of the time, he could only achieve 108 hours with a strong chance that many of these hours would be in the middle of the night. Wind power was free, so the majority of mills used it whenever possible, reverting to steam only in long periods of calm or local emergencies. Most engines were fixed in the adjoining shed, but some were portable and some mobile so that they could be brought to the mill only when needed.

When using steam power, the speed of the stones could be better controlled and run at steady higher speeds than wind could reliably provide. 130 to 140rpm could be held for long periods to give a finer flour that, with the high-speed graders and cleaners, produced a near-white flour. With these quality aspects and reliable production, some mills installed steam engines and boilers to drive all the stones and flour dressers in the mill and this needed 16 to 20hp. Such an installation needed a large investment and the boilers required stacks that were high – often reaching 20 to 30m – much higher than the mill itself. Continuing this trend, as gas and oil engines became available, the steam engines themselves were replaced to keep stone milling going.

CHAPTER 10

The Milling Process

It was fortunate for man that the extraction of flour from grain seemed an uncomplicated process. It could be done by crushing between two stones moving in a vertical direction, or rubbing backwards and forwards in a horizontal plane. From the latter, continuous forward motion between two stones became the small hand-turned querns and true milling was born. The actual process of milling could be well controlled by man's own senses. Touch by his thumb indicated fineness and temperature and 'miller's thumb', 'rule of thumb' and 'golden thumb' are expressions that have become folklore from such control. By smell he could detect smut, mould and rancidity. He had no instruments so he had to rely on his natural senses and experience, and for these reasons there are some wide and interesting variations on how good meal can be achieved via what proved to be a very complex process.

The Stones

Diameter, speed, rotational direction, power available, type and hardness, furrow pattern, gap between and the grain-feed rate all affect milling. The maximum power available is fixed before building by the decisions on mill type, height and sail area, and can only be increased by major rebuilding. The milling of flour developed from the small-diameter, smooth-faced querns that were turned by hand at 30 to 40rpm. Then came the beast and vertical watermills, where larger stones could be turned for longer periods but still at low rotational speeds as they were directly driven. Horizontal-wheeled watermills with one-stage gearing permitted higher speeds, followed by two-stage gearing that allowed speeds of over 100rpm to be reached. The post mill is a one-stage and the tower mill a two-stage unit where the same speeds can be achieved. With varying wind speeds, the milling range is between 80 and 120rpm.

Diameter

Typical stone diameters, speeds, and horse power as the milling equipment evolved are given below.

Gearing	Mill type	Stone Diameter (ft and m)	Rotation (rpm)	Rim Speed (mph)	Power (hp)
Direct	Hand quern	1.5; 0.4	30	1	0.1
	Beast mill	3.5; 0.9	10	1	1
	Vertical windmill	7; 2.1	30	10	2–3
One-stage	Vertical water or Post mill	4; 1.2	80	14	3–4
Two-stage	Horizontal water or Tower mill	4; 1.2	120	17	3–4
	Steam	4; 1.2	140	20	4–5

It should be noted that there are wide variations about these numbers, so they have been rounded to show the trends that took almost 2,000 years to complete. Naturally, when man had only small slow hand querns to make his flour, he dreamed of bigger and faster stones to make more flour, and larger stones were installed to achieve this. The principal is sound when starting from small stones, but high speeds in large stones (say 2.1m diameter) give very high rim speeds that produce fine, dusty and over-heated flour that soon goes off and makes poor bread. It is also much more difficult to quarry and handle a 2.1m-diameter stone weighing about 3 tons than a 1.2m-diameter stone weighing only 1 ton.

In the Peak District of Derbyshire, there is an area of the larger abandoned stones overgrown by mountain oaks that would indicate that the production of such stones stopped at least before 1800 and probably much earlier. Of the 400 stones found most were 1.8 to 2.1m diameter, with almost half of them rejected when partially completed, confirming the difficulty of manufacture of large stones which adds to the cost. It is likely that a 2.1m-diameter stone cost three to four times more than a 1.2m-diameter stone.

There had to be a compromise to get good flour at a high production rate and acceptable cost, and from about 1500 it began to be realised that medium-diameter stones run at higher speeds gave the best solution, but it would as late as 1800 before this solution became standard practice. The changes in rim speed by altering diameter or rotational speed are shown on the following page.

Increase in rim speed with diameter at 100rpm:

| | Diameter | | Rim Speed |
	ft	m	mph
Eye	0.75	0.23	2.6
Rim	3.5	1.05	12.2
	4	1.2	13.9
	4.5	1.4	15.7
	5	1.5	17.5

Increase in rim speed with rotation at 4ft diameter (1.2m):

Rotation, rpm	Rim Speed, mph
80	11.2
100	13.9
120	16.8
140	19.6

The final consensus was stones between 1.06 and 1.5m diameter with more than 80 per cent being 1.2m rotating at 100 to 120rpm with rim speeds of 14 to 17mph.

Stone Type

Up to the early nineteenth century, any local hard stone was used for both flour and animal feed, and these stones were quarried in one piece. The most well known of these grey or grit stones came from the Peak District of Derbyshire – hence their more usual name of Peak stones. The main quarries were in the Hathersage area, where the millstone grit contained tiny quartz pebbles that aided milling. Hexagonal blocks about 350mm thick were cut from the quarry face and inspected for flaws before any further work was done. They were raised off the floor on pedestals to a convenient height for working and trimmed circular and the top flattened. Once turned over, the other side was flattened and a square or round hole cut in the centre. At any stage, a crack, plane of weakness or soft area would mean rejection of the stone, but it was the skill and experience of the quarrymen in selecting the location of the initial block that kept such rejects to a minimum, probably less than 10 per cent. These 1-ton stones were dragged, slid or rolled by horse to the nearest road and transferred to horse-drawn carts. Their quality was such that it was worth transporting them along the toll turnpike roads or canals to most parts of England. About 1800, a pair of 1.2m-diameter Peak stones would cost about £5 at the quarry.

Transport and several tolls at 2s 6d a time would add significantly to their delivered cost, but it was worth paying. Even when their use for flour ceased, their continued use for animal feed meant that some 20–30,000 were still being used in the 1880s, with an annual production of about 3,000 for windmills alone.

After the French Wars ceased in 1815, the European practice of using French burr stones, which lasted longer than Peak stones, was taken up, and burr stones began to be imported for English mills. The burr is a very hard quartzite and the best comes from La Ferte-sous-Jouarre, which is about 50km east of Paris near the River Marne, where seams are no more than 300mm thick. It is therefore rare to find a lump of burr big enough to make a one-piece mill stone, but Heckington Mill in Lincolnshire has such a stone. The burr stones have, therefore, to be made from several lumps (Figure 45).

The pieces for the milling face were selected from lumps that already had a near-flat face. Such lumps were then cut at right angles to the flat face to give an edge of about 150mm thick or 6in. A wooden former the diameter of the required finished stone was set on a smooth level surface and pieces of burr individually chiselled to fit the outer rim. The craftsmen selected the pieces nearest to the shape needed so there was the minimum of chiselling for the arc and the other edges were made straight, at right angles to the face and sides, to make matching easier for subsequent pieces with the minimum of gap between them. The selection and chiselling continued to the centre of the stone where a square hole was made for the bed stone and a round one for the runner. Sometimes the central part with the hole was cut from a piece of Peak stone and the outer area only made from the more expensive burr. Some manufacturers went even further and chose large pieces which were cut as segments of a circle, to reduce the number of pieces used and give a degree of symmetry. The burr stone face was therefore made from 1 to 40 bits, but 12 to 25 in a random pattern was more usual. Random pieces of burr were added to the back and cemented together with plaster of Paris to make the stone about 300mm thick (1ft). The whole block was bound round with two or more iron bands to prevent bursting of the stones at operational speeds.

Dressing the Stones

Quartz, quartzite and millstone grit are very hard minerals that need even harder tools to cut and dress them to a standard of precision that will permit a 1-ton stone to rotate at 120rpm without wobble and have patterns cut into the face to grind the grains into flour. Such tools are very special!

The Dressing Tools

The special steel chisel is called 'the bill' and its wooden handle, 'the thrift' (Figure 46 and Colour 42). Wrought iron is too soft to cut stones but the technique of hardening this material evolved to make weapons like swords and spearheads. It became known that some irons were more difficult to work when cold than wrought iron, and when such irons are heated to a high temperature and plunged into water a very hard steel is produced. Over the centuries, the blacksmiths learned the steps to make the best bills. A tough, very hard to bend when cold, steel was selected. It was hammered into a double wedge about 40mm² in the middle and 300mm long. One tip was heated in the charcoal fire to bright or cherry red and then quenched into water until black and cold. The other tip was similarly treated, but this time the whole bill was immersed in the water and left to go cold.

The metallurgy of the hardening process which was, of course, unknown to the blacksmith is as follows:

In selecting a tough iron he was in fact selecting an iron containing more carbon, which is now known as steel. Wrought iron has less than 0.1 per cent carbon whereas the tougher steel has 0.4 per cent carbon and a very hard steel would have about 0.7 per cent carbon. Heating this last steel to cherry red heat gets the steel to 750–800°C where the microscopic structure changes and quenching this into water produces a very, very hard but brittle steel. However, if the steel is reheated and then slowly cooled, it will revert to its normal room-temperature softer state so that the smith knew he had to keep the bill tip in water till cold so heat from the thick middle section could not conduct along to the tip to soften it.

It is a great compliment to the blacksmith that, without chemical analysis or temperature measurement, he was able to produce bills that would cut and shape quartz and millstone grit. The point was ground to 40 degrees from both or one side only and the bill was finished. However, being as hard as it was, it could not be pierced and mounted on a shaft but had to be held so that accurate cutting could be done. This problem was solved by the thrift, a hard wood handle with a larger diameter barrel at one end with a slot giving plenty of clearance for the bill. The bill was wedged into this slot by thick pads of horse-harness leather and was then ready for use. (In the latter part of the nineteenth century, as metallurgical processes improved, pierced bills with hammer shafts could be made, but they remained a rarity.)

In use, the very hard, brittle and heavy steel bill hits an almost equally hard stone. The bill, therefore, does not wear gradually and uniformly but looses small flakes and chips that fly off and roughen and break the cutting edge. These chips embed themselves in the backs of the hands of the dressers and, being iron, discolour the skin with reddish-purple flecks which were recognised as evidence of the dressers experience. Millers would demand that the dressers 'show their metal' as proof of their skill before employing them. The chipping took the edge off the bill in as little as half an hour, and since it would take half a day to dress a stone, 8 to 10 would be used and they would be frequently ground for further use. A new bill would be 300mm long and weigh 1.5kg with a 50mm hardened tip. The regrinding process reduces the weight of the bill, so it was general practice to use new heavy bills for furrow cutting and older lighter ones for the more delicate job of cutting the fine lines on the lands. A bill would not be discarded until it was as short as 150mm, weighing only 0.75kg.

The dressers other important piece of equipment was the 'paint staff'. This was a narrow board of hard wood 1.25m long, one side of which had to be truly flat. So that it kept its flatness and straightness, it was made from three laths cut from one plank of straight grained wood restuck together with the centre lath reversed to almost eliminate warping. The flatness on the long side was checked against a 'proof staff' made of cast iron, slate or steel, kept polished and lightly oiled in its own substantial box for protection. The paint staff was laid on the proof staff and any high areas would be oiled. These were removed by stone or steel scrapers and the paint staff rechecked until it was truly flat. It was then daubed with a mixture of red iron oxide and tallow and moved gently across the face of the mill stone. Any high spots were coloured red and these were carefully removed with the bill or a burr rubber and the process repeated until the face was absolutely flat. The truly flat bed stone was now ready for the marking out and cutting of the furrows, while the runner stone had to be very slightly dished or swallowed (Figure 47). This concavity of the stone was about 3mm but it is critical for good milling, though no particular tools seem to have been used to make it. Perhaps one side of the paint staff was profiled, or maybe the dresser did it by eye and experience.

He also had furrow and land sticks which were hard wood laths about 1m long, 6mm thick and the width of a furrow or land. Furrow sticks were 30 to 40mm wide and land sticks 40 to 100mm wide. They were kept in pairs where each pair would give a particular pattern of furrows and lands on a stone face which would be determined by the mill stone maker. Redressing would follow the original dress and it was the mill that kept the tools for the itinerant dressers to use.

The Furrow Patterns

These are very diverse and their interaction in the milling process is difficult to visualise. Furrows began to be cut into the smooth faces of hand querns, possibly very soon after the quern itself had been developed. Whenever they first appeared, they were in widespread use by Roman times – a millennium before windmills appeared. The first simple pattern was probably four straight radial symmetrical furrows that divided the stone into four quarters, but trial and error soon increased their number, giving more sectors to get better milling. The name quarter is retained to this day even when seven to eleven sectors are used. Their more common name is, however, harp, possibly because one sector with its secondary grooves does resemble the musical harp, and these are the most common for flour and feed milling in England. An alternative is the use of many curved furrows without secondary furrows, referred to as a sickle dress because the curve resembles the blade of the hand crop cutter. This was preferred by the Dutch but did find some acceptance in England.

Stones rotate clockwise or anticlockwise depending on the selected rotation of the sails and type of gearing used in the mill. On both post and tower mills the sails almost always rotate in the anticlockwise direction, but the post mill is a one-step

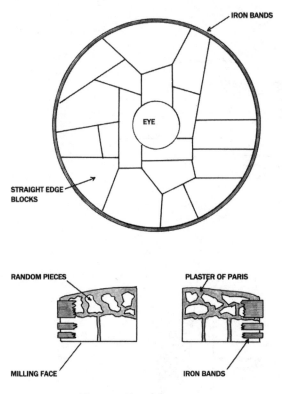

Figure 45. French burr stones.

gearing system and the tower mill two. The stones in a post mill therefore rotate anticlockwise, and in a tower mill clockwise. As always, in a few cases the opposite applies but using the ratio of post mills to tower mills and the higher number of stones in the tower mill, there would be more clockwise stones than anticlockwise. A dedicated researcher of millstone dress, on a sample of sixteen stones, found six clockwise, seven anticlockwise and, very significantly, three with radial dress. A guess, rather than a scientific estimate, is that in England 50 per cent of stones were clockwise, 35 per cent anticlockwise, 10 per cent sickle and 5 per cent radial. The three predominant types are shown in Figure 48, which clearly shows the offset of the main furrows. The draft circle to produce this effect ranges from 50 to 400mm and is done to make the furrows intersect at an oblique angle during milling. The curve of the furrows on the sickle dressed stones has the same effect (but on radial stones the edges meet along their whole length but do not cross). The purpose of the intersecting is to cut the grain and help strip the bran from the berry. The angle of the furrow edges also helps to push the meal to the rim of the stones.

On the prepared surface of the stone the required pattern of furrows was marked out using the pair of furrow and land sticks. The circumference of the eye was divided into the number of harps needed – seven to eleven – and marked. The furrow stick

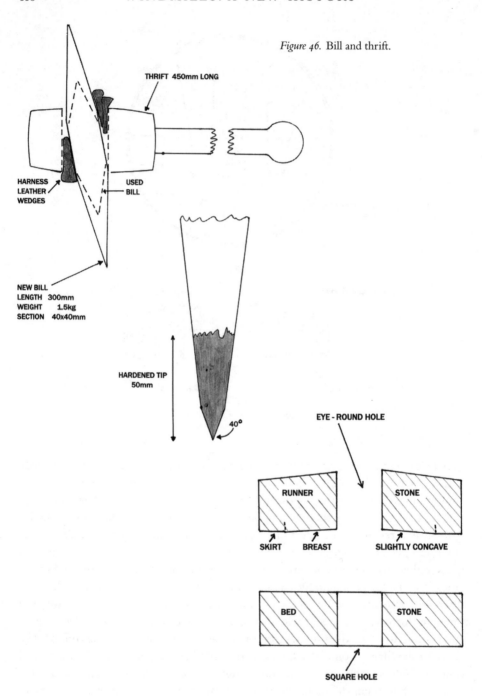

Figure 46. Bill and thrift.

THRIFT 450mm LONG

HARNESS
LEATHER
WEDGES

USED
BILL

NEW BILL
LENGTH   300mm
WEIGHT   1.5kg
SECTION   40x40mm

HARDENED TIP
50mm

40°

EYE - ROUND HOLE

RUNNER

STONE

SKIRT   BREAST

SLIGHTLY CONCAVE

BED

STONE

SQUARE HOLE

ALTERNATIVE PREPARATION OF RUNNER STONE

SLIGHTLY CONCAVE FROM EYE TO RIM

PERFECTLY FLAT, LIKE BED STONE

Figure 47. Stone preparation.

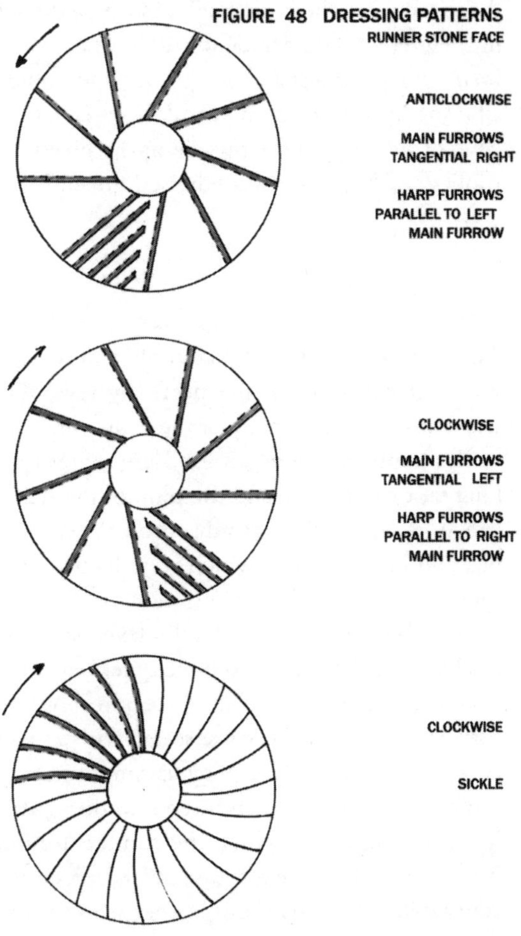

Figure 48. Dressing patterns.

FIGURE 48   DRESSING PATTERNS
RUNNER STONE FACE

ANTICLOCKWISE

MAIN FURROWS
TANGENTIAL RIGHT

HARP FURROWS
PARALLEL TO LEFT
MAIN FURROW

CLOCKWISE

MAIN FURROWS
TANGENTIAL LEFT

HARP FURROWS
PARALLEL TO RIGHT
MAIN FURROW

CLOCKWISE

SICKLE

was used first and positioned on a mark and angled to give the wanted offset from radial at the centre. The master furrows were then traced on the stone by the edge of the bill or a brush dipped in reddened tallow. Next, the paired land stick was used to mark the ends of the secondary or harp furrows and from the other furrow, the land and furrow sticks marked the position of the shorter harp furrows (Figure 49). This was repeated for the remaining harps. It was a time-consuming and skilful task, and some of the larger mill stone companies had wooden formers made that fitted over the whole stone with the complete pattern of furrows and lands that could rapidly, reproducibly and repeatedly mark many stones.

At the makers, the master and harp furrows were then cut with the heavier bills. Again, even before windmills existed, the profile of the furrow had evolved to a shape that more efficiently cut and moved the grain and meal. The furrow had a near-vertical edge 12 to 15mm deep and a shallow slope of about 30 degrees that merged into the land (Figure 50). The edge was variously called black edge, back edge, cutting edge or sharp edge, and the slope feather or land edge. In Figures 49 and 50, the sharp edge is shown as a full line, and where slope and land merge is shown as a broken line. Once the furrows were cut, the stone was ready for the final operation of chiselling the stitches on the lands that completed the fine grinding of the grain fragments into meal.

The smooth lands were 'feathered', 'cracked' or 'stitched' with fine lines cut by the bill to give a file or rasp-like surface for grinding the flour. For the most part the stitches were parallel to the furrows, but a significant minority had stitching at 45 degrees to the furrows (Figure 49). The latter was more common where higher speeds could be maintained on the stone, and for steam-driven mills operating at 140rpm, 45 degrees was standard. Traditionally, the stone makers for windmills left the stitching to the individual mills to be done by the travelling dressers or the miller himself. He would decide on the number of stitches, which was typically eight per inch — a separation of about 3mm — but could be as many as eighteen per inch, and all done by eye, steady hand and experience. The stitches extended from the outer edge of the skirt to the middle of the breast and whichever dress was selected, both runner and bed stone were dressed the same.

Setting the Stones

The bed stone had the square hole and was laid on the substantial oak stone beams where it had to be made perfectly level. This was done using three wedges at 120 degrees separation under the stone. Since levels had not been invented, a specially made plumb line like an inverted large letter T was used. The horizontal bar about 1.2m long sat on the stone and the plumb line was attached to the top of the 1m vertical lath. When the line indicated vertical, the stone was horizontal. This was checked at the position of the wedges which were tapped in or out to adjust. When the plumb line showed vertical at all three positions, the bed stone was truly horizontal. The wedges were pegged or nailed to lock this level position. Once screw threads became available, levelling could be achieved more quickly and held by lock nuts on the levelling screws. The bed stone protruded 50 to 100mm above the planks of the stone floor and the gap between stone and plank was sealed with a small lath bent round the stone and nailed. Any residual gap is plugged by hammering in a length of cord.

The neck box was then fitted into the square hole of the bed stone. The neck box held the spindle on which the runner stone rotates. From simple wooden blocks, it eventually became a sophisticated piece of engineering in its own right, using adjustable brass collars and leather seals to hold the spindle vertical and permit up and down movement of it while rotating at up to 140rpm. The seals prevented dust getting between the spindle and the bearing surfaces which would cause severe wear. The spindle must be set vertical and this was done using the neck box and the bearing block located on the bridge tree under the bed stone. The spindle was slid into its working position through the neck box collars and into the bearing block which had four horizontal screws for adjustment. The spindle, 500 to 600mm long, with a square section at the top onto which was fixed the jack staff. This was another simple but effective tool that ensured that the spindle was truly vertical. It was a lath of hard wood about 1m long with a square hole at one end that slid loosely over the square section of the spindle to which it could be tightly wedged so that the lath was about 150mm above the bed stone. At the outer end of the jack staff, a long goose feather or sliver of wood was pushed into a small hole so that the tip just touched the surface of the bed stone about 50mm from the edge. The jack staff was rotated so the feather moved round the outer surface. The feather may have bent or left the surface of the stone, both of which indicated that the spindle was not vertical. The screws in the bearing block were adjusted and the slackened collars in the neck box allowed the spindle to move slightly until full rotation of the jack staff showed no change in the tip of the feather relative to the stone. The spindle is then truly vertical and is locked in position.

The runner stone is carefully lowered into position with the domed top of the spindle engaging the cupped hole in the rynd (Figure 51). The spindle is in the raised position and the runner stone will rotate freely and easily on it and must be accurately balanced. Pieces of lead or iron were put on the rim of the stone and moved by trial and error until static balance was achieved. These pieces were then fixed to the stone to retain the balance so the stone would run true. This was a very critical operation as the 1-ton runner stone rotated at up to 140rpm with outer rims only the thickness of tissue paper apart. Any wobble would cause touching and rapid wear of the stone and sparks that could ignite powdered flour and cause fires. A stone trued at rest may not actually be true when rotating, so in 1859 the Victorian engineers Clarke and Durham patented a system to cure this problem. Four pockets, 100mm deep and 50mm diameter, were cut into the stones and were fitted with central screws. Large round iron nuts of various thickness were screwed on to get the 'at rest' balance position. The stone was then rotated to see if it wobbled or dipped. If it did, the weight opposite the dip was screwed up and the process repeated until a proper dynamic balance was reached. The screws could be easily reset when wear changed the balance position. An alternative was to have four horizontal slots equally spaced round the circumference of the stone, fitted with screws that could be turned to move threaded weights in or out until balance was reached. The runner stone could be safely rotated at high speed without fear of contact with the bed stone.

The Grain Feed

The system is essentially the same for both underdrift and ovedrift drive, but because overdrift is more usual the explanation that follows is for that system (Figure 51 and Colour 43). As the stones need frequent dressing requiring the removal of the runner stone, the grain-feed system has to be easily taken apart. The vat or box which contains the meal emerging from the stones gap is therefore made in two halves that are clipped or wedged together, and when separated can be completely removed. In its normal working position, the quant and its crotch that drive the runner stone are put in position and the grain-feed system added on its support frame – called the horse – which sits on top of the vat. The grain hopper and a feed slide – called a shoe or slipper – which directs the grain into the eye of the runner stone, are fixed onto the horse. At the back end of the hopper is a pivot so that hopper and its shoe can move from side to side. The quant has a square section that knocks the shoe four times with each revolution, so shaking the grain into the eye. The shoe is held in contact with the quant by a short rope tensioned by 'the miller's willow'. This is a wooden arm that acts like a spring.

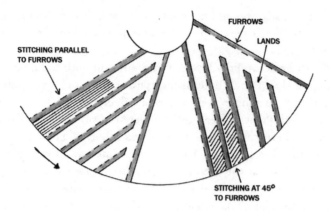

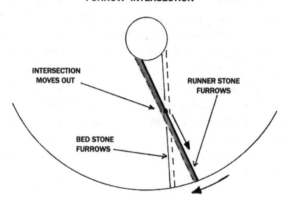

Figure 49. Stitching and furrow intersection.

In underdrift stones, a small shaft extends from the rynd to above the level of the eye. It may be a square shaft or a shaft with three or four loops or bars of iron fixed to it to shake the shoe. At 120rpm, there are 300 to 400 raps a minute, making a continuous noise said to mimic the incessant chatter of women or damsels – hence it is called a damsel. As usual, it is not known exactly when this unit was invented, but its name implies that it was a long time ago, probably on watermills before windmills existed. It is also said that the name – damsel – should only be applied to the rapper on underdrift stones, but the term is now often used for the square section of the quant which, after all, makes the same noise. Whatever the name, it was a very significant development. As a strengthening wind increased sail and stone speed, the number of raps from the damsel increased, thus shaking more grain into the eye. Grain feed is therefore linked directly to stone speed and this may be the first example of proportional control in engineering – long before the Industrial Revolution of the eighteenth century.

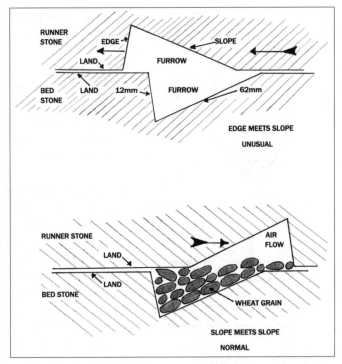

Figure 50. Furrow profile.

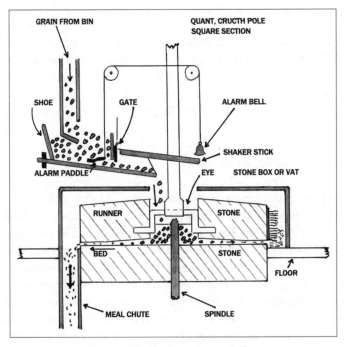

Figure 51. The feed system, overdrift.

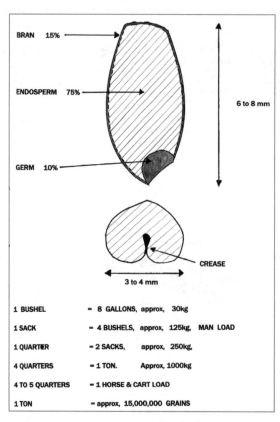

Figure 52. Wheat grain.

1 BUSHEL	= 8 GALLONS, approx, 30kg	
1 SACK	= 4 BUSHELS, approx, 125kg,	MAN LOAD
1 QUARTER	= 2 SACKS, approx, 250kg,	
4 QUARTERS	= 1 TON. Approx, 1000kg	
4 TO 5 QUARTERS	= 1 HORSE & CART LOAD	
1 TON	= approx, 15,000,000 GRAINS	

The other control of feed is the lifting or lowering of the vertical gate on the hopper which can be moved by the miller on the floor below using the crook string. An alternative to this is the same string controlling the slope of a tilting shoe. Another simple but ingenious device is the alarm bell that told the miller that the grain in the hopper was running low. A leather paddle was buried near the bottom of the hopper and held down by the weight of the grain above it. A string from the paddle suspends a bell above the damsel, and when the grain holding the paddle down has gone it is released and drops the bell into the path of the damsel, making it ring until the miller comes to refill the grain in the feed bin and reset the paddle. The bell can be dropped onto any part of the mill's rotating machinery to give the same warning. To ensure that the feed hopper does not overflow, the chute from the grain bin extends below the level of the rim, which restricts the flow of grain into the hopper to the flow that is feeding the stones.

The Action of the Stones on the Grain

It is the purpose of the furrows and lands, by their relative movement between the runner and bed stones, to mill or grind the wheat berry into meal. Figure 52 shows the size and structure of the berry that has to be processed by the stones. There are approximately 15 million grains of wheat in a ton and the furrows distribute them into the lands where fine milling occurs. The furrows are triangular in section, as shown in Figure 50, and a minority of stones are so dressed that the steep edge of the furrows on the runner meets the stationary slope of the furrows on the bed. It is much more common to have the stones dressed and set so that the slope on the runner meets the slope on the bed. The geometric details of furrows on a typical English stone relative to berry size are given on the following page:

Stone diameter, m	1.2
Harps	9
No. of furrows	45
Total furrow length, m	13.5
Cross section, mm^2	360
Total volume, m^3	0.005
No. of grains in bed furrows	70,000

The identical patterns of furrows, as seen when the pair of stones lie side by side, criss-cross one another as the runner stone revolves over the stationary bed stone and the point of intersection of the furrows moves outward for both straight and sickle furrows, just as the blades of a pair of scissors do when cutting material (Figure 49). Sickle dress has twenty-three to twenty-nine furrows and a nine harp straight dress has forty-five, each of which slice across all forty-five furrows on the other stone in one revolution, i.e. over 200,000 slicings in a minute at 100rpm. The incoming grain first fills the furrows in the bed stone, and once the level of the grain is above the surface of the stone the slicing action begins to cut the berry into pieces, the smaller of which are pushed between the lands for grinding even smaller. At the oblique angle that the furrows meet, the major force is to cause slicing and pushing of the sliced pieces into the land gap which causes even the 1-ton runner stone to lift slightly. A smaller force sends the grains tumbling outward along the furrow. The greater the angle of the furrow to the radius, the greater this force. When a land is above a furrow, the rasp-like surface both strips off the bran and cuts fine pieces off the endosperm as a file does to metal. The strong rasping action also rolls the grains and brings grains from the bottom of the furrow to the top, so that new surfaces are exposed for bran removal and grinding. As the grains move outwards, the furrows become shallower to help in bringing the remaining larger pieces to the surface for grinding and contain an increasing amount of flour and bran. By the time the rim is reached, there should be no large bits and only meal emerging from the stones.

The outward movement of the meal is helped by a small induced flow of air from the eye to the rim in the furrows on the revolving runner stone. As the mixture of flour, cut grains and bran tumbles in the air flow, the larger pieces fall back to be further cut and ground, while most of the fines are moved outward along the furrows to reduce over grinding in the lands that produces very fine dust. To address this problem, bottom-driven stones were tried so that the centrifugal force moved the mixture along the lower furrows, but there is little evidence of wide adoption of this system. In the 1840s, several patents were obtained for increasing the air flow through the stones by fans, as this had the added benefit of cooling the meal.

George Hinton Bovill obtained patents in both Scotland and England in 1846 and 1849 and persuaded the Admiralty to let him equip, at his own expense, three pairs of stones at their Deptford Victualling Yard mills, so the process could be properly evaluated. He met considerable opposition from the millers at Deptford, who did their best to sabotage the trials and prove the process a failure. Bovill complained and a Board of Enquiry was set up by the Admiralty to investigate. Parliament ordered the publication, in April 1853, of all the test results and letters. Bovill was vindicated, the head miller at Deptford was dismissed and the process was adopted for six pairs of stones at Deptford only. No other Admiralty yards were equipped, but Bovill had the proof he needed to validate his patents and market his process to the rest of the milling industry. The trials showed how effective the correct use of air could be and the benefits are summarised below.

1.   The milling rate doubled.
2.   The meal temperature was reduced by up to 20 degrees farenheit so it could be dressed directly from the meal chute.
3.   The quality of the flour was much better – over grinding had been reduced so there were less fines and the lower temperature avoided rancidity.
4.   The stones milled up to four times the weight of grain that unventilated stones did before re-dressing was needed.
5.   With the inflow of draft to the eye balanced to the exhaust into dust chambers, dust in the mill was just about eliminated.

There were imitators of Bovill's system so he took them to court for infringement of his patents. Out of twelve cases up to 1868 he won eleven, and was awarded costs and the right to have royalties of 1*d* to 3*d* per quarter of grain milled backdated. The process was widely used by the big steam mills with up to fifty pairs of stones, and to a lesser extent by smaller watermills, making Bovill a multimillionaire in today's money. Windmills do not seem to have taken up the process, probably because they had no surplus power for fans, and fans need to operate at their design speed to maintain air flow which is difficult in varying winds. Furthermore, the traditional wind miller had neither the money to fund the system nor the engineering skills to control and maintain it, so they continued with their age-old process taking its time to make flour. For them, a pair of 1.2m-diameter stones turning at 120rpm in a good wind would make about 150kg of meal per hour or 2.5kg per minute. Between the stones was about 3kg of grain so it would take about 1 minute for a grain entering the eye to emerge as meal from the spout. It would have been accompanied by about 70,000 others.

Grain Cleaning

The stones will mill all that is fed into them, so the quality of the meal was very dependent on the condition of the grain, which meant careful attention to harvesting and storage. The corn was harvested by sickle or scythe and bound into sheaves. Horse-driven self-binders were introduced in the early 1800s and used for a century or more. The wheat had straw 600 to 700mm long, which is much longer than today's wheat of 450mm to suit the combine harvester. Timing of the harvest is always critical and weather dependent. Wheat should be harvested just before being fully ripe, or the bran on the berry becomes thicker and reduces flour yield. The ear must be yellow, but the straw just below it should have no sap when crushed between the fingers. The grain should be as dry as the weather permits when cut. Ten to twelve sheaves were stacked the same day as they were cut, and then turned every following day to help with the drying. Within a couple of weeks, the sheaves were carted to near the farmyard where rectangular ricks, made of layers of sheaves, were topped off with a house-like ridge. The ricks were big – holding sheaves from 10 to 15 acres – and the final operation was to thatch the ridge with wheat straw to keep out the rain.

Harvesting lasted about six weeks and some grain was threshed at once for the mill. Most of it remained in the ricks to be threshed in autumn or spring, as more flour was needed. The ricks were sized so that they could be threshed in one day to give 50 to 60 quarters (12 to 15 tons) of grain. Separating the grain from the straw was called 'winowing' i.e. blowing with the wind or fans. However, there was still some straw, dust, grit and small seeds that went in with the grain for the mill. There was also a wide range of berry sizes that did not help with consistent grinding, and smut – a form of mould on the berry – could contaminate the meal.

Dry grain should have no more than 16 per cent moisture and could be stored in sacks, in deep heaps mixed with chaff in closed dark granaries or in shallow beds with daylight where regular turning was necessary. Wet grain from bad weather at harvest or poor ricking could have as much as 30 per cent moisture. Such grain will sprout in the ricks or warm up in the granaries after threshing and go mouldy. The old way of drying was to spread the grain on the upper floor of the granary to about 150mm deep and turn daily to get the moisture out. Some granaries added a stove on the ground floor, heating air that passed through small holes in tiles laid on the upper floor to drive off the wetness. They had two practical ways of deciding if the grain was dry enough to mill. One was to press a ball of grain into the palm of the hand and see if it remained as a ball. The other was to try to push the hand into the grain and see how far it went in easily. If the ball fell apart or the hand went in up to the elbow, the grain was right for milling. The dry grain was only bagged up when it was about to be sent to the mill.

The dry grain at about 16 per cent moisture, if it could have been measured, mills well and cool and gives a good meal that will, itself, store for a few weeks. At 17 per cent moisture, the meal warms up but is still good. At 18 per cent, more power is needed to mill it so the meal warms up more, may have a slightly singed taste and tends to go off more quickly. Above this, the grain can stick in the hoppers and bins and the meal gets very hot during milling, so much so that the enzymes in the berry break down and give a rancid taste. Also, with the moisture, the meal makes a stiff dough that clogs up the furrows and stops the milling process. The stones can be cleared in the early stages if a batch of hard dry grain is run through, with the stones lifted so the unbroken grains scour out the clogged furrows. If the dough gets really hot and crusts up, the stones would need to be lifted out and scraped clean. The miller had therefore to be very careful how he dealt with the grain that was delivered to his mill, and no doubt there must have been several lots rejected. Even the accepted grain would contain dust, mould, straw bits and seed, and for several hundred years the flour the customers got was only as good as the grain they brought to the mill. It was the baker who had to sift by hand to remove such trash to get clean flour for bread making.

The oldest form of grain cleaners were bigger versions of hand sieves, shaken or rattled by a drive from the main shaft on the bin floor or by beast power in a ground-level shed. It was known as a jogscry and later models with two or three sizes of sieve could not only clean but also separate the grain into two sizes to be ground separately at different stone settings to get a bigger yield of flour. Its efficiency was poor, so improvements were really needed. So, in 1768, Robert Mackell and Andrew Meikle (yes – the inventor of the spring sail) took out patent No.896 – 'a machine for dressing wheat, malt and other grains before they are ground, and effectively cleansing them from sand, dust and smut and consequently greatly improves the flour product from wheat so dressed'. The main change was a rotating, inclined, cylindrical sieve section 1 or 2m long to get better cleaning, and it was hence unlikely to fit into the dust floor of a mill and would have been installed in a barn and powered by a horse or cow. The cleaned grain would be sacked and hoisted to the top of the mill.

Hand threshing was both slow and inefficient, and in 1788 the fertile brain of Andrew Meikle conceived the first mechanical threshing machine. This is described in patent No.1,645 and used cow power to thresh and get reasonable removal of straw to get a better grain for milling straight from the threshing machine. The lower part of both of Meikle's inventions was an open winnower, so in 1800 Cooch brought out a closed winower with integral sieves that was so successful that it was still being used in the early 1900s. From the smaller farms, uncleaned grain was still being sent, and the receiving mills installed their own grain cleaners on the dust floor driven at about 20rpm from the main shaft. The dust, seeds, sand and smut were almost completely removed, as were small grains of wheat so that only the large grains were milled for flour (Figure 53).

The Miller's Control

The Peak stones were used for oats, rye and beans to provide animal feed, as such stones were not considered hard enough to produce a good fine flour from the harder wheat grains but, before the introduction of burr stones, they were used to make a coarse wholemeal flour. Once steam power had been introduced, burr was used for flour production on stones which could operate at a steady 140rpm, but the wind miller had to make flour in variable winds and hence at varying stone speed. Reefing of common sails helped to reduce the speed range and spring and patent sails further improved the level of control, but still did not produce a constant speed. Milling therefore occurred between 80 and 120rpm, rarely holding constant for more than a few minutes. The miller and the machinery had to deal with this by regulating the grain feed to the stones and the gap between them to get a good, fine, consistent flour.

The miller assessed the quality of the meal as it emerged from the chute to be collected in sacks. He took a sample in the palm of his hand and rubbed this with his thumb to judge the fineness, texture and the temperature. The dependence of good flour on the miller's thumb gave rise to the sayings rule of thumb and golden thumb, to mean a practical test for assessing any process. The delicate aroma from the warm meal also had to have no hint of rancidity. To get all these parameters right as the stone speed varied with the wind, the miller had two controls:

1. The rate of feed of the grain to the stones, controlled by the crook string.

2. The gap between the stones, controlled by the governor or manually by the tentering screw.

These controls are not totally independent so there is a range of settings that will give a good meal but, conversely, there is a wider range that will make poor meal. Too high a feed rate slows the stones and makes the runner stone lift, so the meal is coarse and cool and may clog the stones or let large bits of grain through into the sack. At a steady feed, too large a gap makes the meal coarse and too small a gap makes the meal dusty, hot and rancid.

The automatic part of feed control is the damsel, which shakes the feed shoe in direct proportion to stone speed, i.e. higher speed causes more feed and vice versa. The manual part of feed control is the setting of the height of the gate on the shoe or its slope by the crook string, which passes over pulleys and down through the stone floor to near the meal chute where the miller stands.

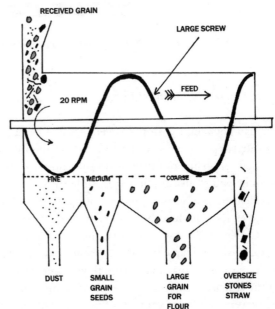

Figure 53. Grain cleaner.

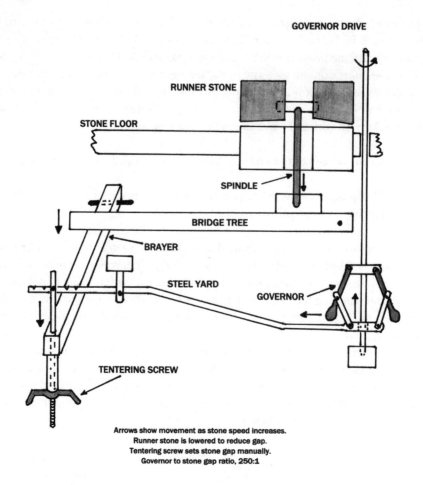

Figure 54. Tentering by the governor.

Stone Gap Adjustment by Hand

The gap between the stones is very small – thousandths of an inch – but must be regulated as stone speed varies to maintain the fineness of the meal. This regulation is called tentering and it requires the runner stone to move up and down while running, and so its spindle has to slide in the neck box bearing in the bed stone. The bearing box of the spindle is supported on the bridge tree, which is pivoted at one end, while the other end rests on another beam called the brayer which also pivots. The oldest way of adjustment was to use a crow bar to the lift the free end of the brayer and tap low-angled wedges into the slot to hold the position. It had to be done so often that a tentering boy was employed full-time to adjust the wedges on the instructions of the miller. Knocking a shallow wedge out by 50mm causes the brayer to drop by 5mm and the brayer end of the bridge tree to drop by 1mm.

C1: Bourn Post Mill, Cambridgeshire.

C2: Wrawby Post Mill with roundhouse, Lincolnshire.

C3: Wicken Village Smock Mill, Cambridgeshire.

C4: Northfield Smock Mill, Soham, Cambridgeshire.

C5: Hewitt's Tower Mill, Heapham, Lincolnshire.

C6: North Leverton Tower Mill, near East Retford, Nottinghamshire.

C7: Mount Pleasant Tower Mill, Kirton Lindsey, Lincolnshire.

C8: Skidby Tower Mill, near Hull, Yorkshire.

C9: Burgh Le Marsh
Tower Mill, near
Skegness, Lincolnshire.

C10: Heckington
Tower Mill,
Heckington,
Lincolnshire.

C11: (Top left) Ellis Tower Mill, Lincoln.

C12: (Top right) Sibsey Trader Tower Mill, near Boston, Lincolnshire.

C13: (Bottom) Tuxford Tower Mill, Tuxford, Nottinghamshire.

C14: Waltham Tower
Mill, near Grimsby,
Lincolnshire.

C15: Bircham Tower
Mill, near King's Lynn,
Norfolk.

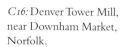

C16: Denver Tower Mill, near Downham Market, Norfolk.

C17: Mediterranean–type Tower Mill, Fuerteventura, Canary Islands.

C18: Thurne Dyke Wind Pump, Thurne, Norfolk.

C19: Horsey Wind Pump, near Stalham, Norfolk.

C20: (Top left) Wicken Fen Wind Pump, Cambridgeshire.

C21: (Top right) Farm Windmill, Fuerteventura, Canary Islands.

C22: (Bottom) Scrooby Sands Wind Farm, near Caister-on-Sea, Norfolk.

C23: Cold Northcott Wind Farm, Cornwall.

C24: Mablethorpe Wind Farm, Mablethorpe, Lincolnshire.

C25: (Top left) Post mill supports, Great Cransden Mill, Cambridgeshire.

C26: (Top right) Bevel gears of fantail to Pinion Drive, Ellis Mill, Lincoln.

C27: (Bottom left) Cap centring wheels and pinion drive on curb rack, Heckington, Linconshire.

C28: (Bottom right) Common sail, Green's Mill, Nottingham.

C29: (Top left) Cubitt's patent sail, North Leverton, Nottinghamshire.

C30: (Top right) Patent sail, shutters and twist, North Leverton, Nottinghamshire.

C31: (Bottom left) Cross-arm sail carrier, Wrawby Post Mill, Lincs.

C32: (Bottom right) Canister sail carrier, Great Cransden Mill, Cambridgeshire.

C33: (Top left) Spider and cranks of patent sail, Foster's Mill, Swaffam Prior,
Cambridgeshire.
C34: (Top right) Weights for patent sail control, Ellis Mill, Lincoln.
C35: (Bottom left) Neck Bearing, Bourn Post Mill, Cambridgeshire.
C36: (Bottom right) Tail Bearing, Bourn Post Mill, Cambridgeshire.

C37: (Top left) Brake wheel – wallower drive, North Leverton, Nottinghamshire.
C38: (Top right) Great spur wheel – stone nut drive, Waltham Mill, Lincolnshire.
C39: (Bottom left) Brake Lever, Ellis Mill, Lincoln.
C40: (Bottom right) Stone drive and groove pattern on bed stone, Burgh le Marsh, Lincolnshire.

C41: (Top left) Sack hoist Y wheel for chain loop, Heckington, Lincolnshire.

C42: (Top right) Thrift and bill for stone dressing, Waltham Mill, Lincolnshire.

C43: (Middle left) Feed shoe, horse, alarm bell and vat, Heckington, Lincolnshire.

C44: (Middle right) Governor for stone gap control, Waltham Mill, Lincolnshire.

C45: (Bottom right) Bridge tree and spindle, Waltham Mill, Lincolnshire.

The spindle and runner stone fall by only 0.2mm so the overall leverage is 250 to 1, giving a fine and accurate gap control.

As soon as threaded rods could be made the wedges were replaced by the tenter screw, and by changing the location of the levers the point of adjustment was moved to be conveniently near to the meal spout where the miller himself could control the stone gap. As wind speed increases the runner stone speed, the increased feed rate makes it lift and increases the gap. This is counteracted by lowering the spindle, decreasing the gap which increases the nip of the stones, producing a braking effect and slowing the stones so flour is milled at a more constant speed.

Automatic Tentering

In 1787, Thomas Mead obtained patent No.1,628 for 'regulating all kinds of machinery where the first power is unequal'. As described earlier, one section of the patent had an idea to control sail area with a ball-operated governor to keep sail and stone speed constant. It did not work, and even the later and more successful spring, roller and patent sails could not keep sail speed constant enough for the milling process. Mead therefore proposed a ball governor to automatically change the stone separation as the speed of the runner stone changed (Colour 44). Included in it was the manual screw adjustment to set or alter the gap and moveable pivot points to change the degree of alteration made by the governor. The system is shown in Figure 54 and detail of the brayer and spindle through the bed stone in Colour 45. It was easy to incorporate into existing tentering gear and was adopted for post, smock and tower mills. The drive for the governor could be aken from the spur wheel where one governor could control one, two, three or four stones with each stone having its own set of tentering levers and screw. With individual governors, the drive is taken from that stone's own spindle. Through the levering system, the movement of the governor collar is reduced 250 to 1. So, when increase of wind speed lifts the governor lever by 25mm, the runner stone is lowered by 0.1mm, which is about one turn of the tentering handle.

It is the skill and experience of the miller that determines the initial settings of the feed gate or shoe slope, and when milling begins and when the stones have reached a steady speed for the wind conditions that day he turns the tentering handle to lower the runner stone from its high out-of-use position, to its working gap. The initial flow of meal is usually directed into an animal-feed sack until the miller is happy that the meal quality is good enough for flour. The damsel and the governor then act automatically to control feed rate and gap to keep the flour quality good, but the miller will not rely on them entirely. He will check frequently by rule of thumb to ensure that the flour is good and adjust the stone gap by the tentering handle to keep it so. This is the primary control, altered several times an hour with the use of the crook string to alter feed rate, changed perhaps once a week.

Separation of the Meal

Whole meal is the complete product of the wheat grain and consists of:

Flour	65 to 70 per cent
Bran	about 15 per cent
Pollards	about 15 per cent
Dust	about 2 per cent

Stone milling is a one-step process, hence all parts of the wheat berry are subjected to the same conditions so that, although the bran and pollards are generally coarser than the flour, some fine bran and pollards are made and will end up in the flour making complete separation impossible. To help the flour separation process, the meal, which is warm from milling, is left to cool in bins before being screened or dressed. The oldest form of dresser was the jogscry, where a downward-sloping chute with medium-sized sieves let the flour and dust through, while the bran and pollards stayed on and fell off at the end. It was shaken or jogged up and down by belt-driven eccentric wheels, but fine bran and pollards still went into the flour. An improved version was invented by Nicholas Boller of Austria in 1502, who added an arm that banged the sack of flour or an inclined tube of sacking through which the meal flowed into the bins. This shook out the dust from the flour.

About 1700 the English Bolter appeared, its name possibly derived from Boller. In this, the meal is fed into a sloping cylindrical frame covered in a cloth sack. It turns at 30 to 50rpm inside a staionary tube fitted with longitudinal bars. The weight of the flour in the sock causes it to sag and bang against the bars, so shaking fine flour through into the outer tube and retaining the coarser flour, bran and pollards in the sock to be discharged from the end. The first cloths were wool, then calico and finally silk. Good separation was achieved and a reasonable fine almost-white flour produced, but the way in which it worked meant that the cloths needed frequent inspection and darning as any hole ruined the separation. The cloth banging on the bars also meant that the speed of operation and throughput was limited, but they were still being used commercially in some of the surviving mills in the late 1920s and a few can, even now, be seen in restored mills.

Probably in the early 1800s, the new form of dresser appeared. In this, the meal was fed into a stationary tubular frame with mesh walls where brushes rotated and pushed the flour through to be collected in three grades, while the coarser bran and pollards went right through and were bagged for animal feed. These dressers could be made longer, tilted at 30 to 45 degrees, and operate at speeds of 300 to 400rpm and so give an excellent separation of the flour at high throughput. These were used mainly in the steam-powered roller mills. Even when wind millers used these machines with auxiliary steam power, some fines from their once through-stone milling ended up in the flour, and it needed only about 0.2 per cent of the darker bran and pollards to make the flour off-white. The home-grown wheats used by the wind miller were soft and chalky, giving a flour that made a dense textured loaf. Imported horn-like grain was too hard to be ground by stones, and flour produced by a five-stage roller plant with inter-stage sieving gave a truly white flour that made a spongier loaf that the public really liked.

Production

The grain was brought to the mill in sacks by cart and the load a horse could move was 1 to 1¼ tons and the sack a man had to carry was 280lb or 127kg. A sack held four bushels of wheat, barley or rye, but the weight was different because a bushel is a measure of volume and not weight. The modern equivalents of the old volumes is given below:

1 bushel of wheat	about 30kg
1 bushel of barley	about 25kg
1 bushel of oats	about 20kg
1 sack or comb	4 bushels – about 120kg of wheat
2 sacks	1 quarter (of a tonne – 250kg)
4 or 5 quarters	1 cart load for a horse

The mill works in wind speeds from 8 to 40mph, but the experienced use of reefing, number of stones engaged and the grain feed to the stones enabled the miller to keep the stone speed for wheat in the range 80 to 120rpm, though hopefully in the better range of 100 to 120rpm. Where production rates can be found there is seldom full data on the number of stones in use, the diameter of the stones or the number of hours operated in the day or week for which output is given. For instance, in the early 1800s, when there was on-going war with France, a detailed Government survey of twenty-five mills was made to estimate England's capacity for flour production. Together, the twenty-five mills made 100 tonnes a week – good enough for information on feeding the people, but not to deduce actual production rates. The range was from 500kg to 7.5t per week, covering post mills with one stone to tower mills with up to four stones, so the specific production rate gives a wide range of 25 to 340kg/hr. Fortunately, there are a few examples of specific information, and two are given below.

1. In a 5mph wind, a 1.2m-diameter stone could only mill 25kg/hr
   When wind speed increased to 20mph, a second stone could be engaged and each stone milled 200kg/hr
2. One, 1.15m-diameter stone milled 135kg/hr
   One, 1.2m-diameter stone milled 180kg/hr
   One, 1.4m-diameter stone milled 225kg/hr

It seems reasonable to conclude that, typically, a single pair of burr stones of 1.2m diameter would mill three to five bushels of wheat per hour, i.e. one sack to give about 85kg of flour. These are hourly rates when the wind is blowing in the right speed range and this, on average, was for only 25 per cent of the time. The wind may not have blown for several days and then blown in the middle of the night, but milling could be done for about 40 hours a week and make about 12 tonnes of flour. Data on the wear of stones and the need to redress is equally varied. Restitching the lands could be required in a couple of weeks or so in a period of sustained high output, or be as much as two months with a more usual interval of one month. Furrows would need recutting about every six months. The wear, of course, related to the amount of flour ground, and it seems on burr stones that restitching would be needed after 15 to 20 tonnes had been milled and refurrowing after 80 to 100 tonnes. Traditionally, the dressing of the stones was done by itinerant dressers who went from mill to mill on a regular round. Restitching probably took half a day, and complete dressing a day, the latter needing thirty to forty resharpening of the bills. The dressing reduced the thickness of the stones, particularly the runner which reduced its weight.

Using the 1.2m-diameter stone as an example, it would begin its life about 300mm thick and weigh 1 tonne. It could be used to about 150mm thick, weighing 0.5 tonnes, but below this it would not have been heavy enough to mill grain. At this point, it could be levelled and used as a bed stone so only a new runner stone had to be bought. The full life of a stone has been quoted as ten to twelve years or as much as sixty to eighty years, so it seems likely that the average life was twenty to twenty-five years. In the 1870s, there were still about 10,000 windmills, plus watermills and steam mills turning 30,000 to 40,000 burr stones and a further 10,000 to 20,000 Peak stones. Annual stone making is estimated at about 10 per cent of capacity i.e., 3,000 to 4,000 burrs and 1,000 to 2,000 peaks. At that time a 1.2m-diameter burr would cost £25 and a peak about £5.

Power

Windmills had been grinding flour for over 500 years and the only concern was the quality and quantity of the flour being produced. About 1750, with the advent of steam power from coal, engineers rather than millers and owners wanted to know how much power was needed to grind flour. But what was power? James Watt, the famous Scottish engineer who made many improvements to the steam engine from 1760 to 1800, measured the work that could be done by a horse. He picked a brewer's dray horse that could pull a 150lb weight via a rope and pulley upwards at 220ft/min. The horse was doing work at 220 x 150ft lb/min or 33,000ft lb/min. In France, about the same time, Desagullers defined manpower as a man raising a hogshead of ale 10ft in a minute. This works out at 6,400ft lb/min or 1/5 of the horsepower. These rates can only be sustained for a few minutes, and a horse working for 8 hours a day can only do so at 22,000ft lb/min and a man, 2750ft lb/min. 33,000ft lb/min became the standard horsepower and was used from that day to compare all power sources. But how does work done by a horse relate to whirling sails? John Smeaton used standard horsepower in his study of windmills in 1759–60 to determine the power generated by his model with 1ft 9in-radius sails. The measured work done by the model sails in a wind speed of 8.7mph was 337ft lb/min or 0.01hp.

Projecting his results to an English mill based on the area of four 30ft-long sails lead him to believe that such a mill was generating 3.7hp. Using modern calculations, the power in the swept circle of 30ft-radius sails at a wind speed of 8.7mph is 12.6hp and at an efficiency of 20 per cent (typical of English windmills), the power to the stones is 2.5hp. Considering the test equipment that Smeaton had, his predictions were extremely good. At the more usual operating wind speed of 15mph, the power for milling is 13hp, or 3.2hp per stone for four pairs of stones, but measuring the power used when milling was impossible then and extremely difficult now. So Smeaton looked at the other ways of milling to obtain confirmation of his data. He found that two horses working in a gin could drive a rape-seed-crushing windmill at 3.5rpm. A windmill with 30ft sails driving a similar seed crusher was able to turn at 7rpm in an 8.9mph wind, which would have been equivalent to four horses or 4hp. Projecting to wind speeds of 15mph gives a mill power of 19hp or 4.8hp per stone. A somewhat higher result, but still remarkably close considering the simplicity of Smeaton's test and measuring devices.

The French engineer, Coulomb, also worked out the power of a rape-seed-crushing windmill in Flanders. He monitored the mill and concluded that the 31ft-radius sails in a wind speed of 13mph were working at 8.2hp and an efficiency of 18 per cent. Correcting for wind speed and sail area gives the power from the mill in a 15mph wind as 12hp, which is in very good agreement with the Smeaton data making about 3hp per pair of stones a reasonable value of power from common cloth sails. Both Smeaton and Coulomb used rape-seed-crushing windmills in their attempts to measure power because the crushing was done by pounders, whose weight and distance raised in a minute can be measured accurately to get the power in ft lb/min and hence the horse power. They logically then deduced that a mill with the same sized sails turned by the same strength wind – but milling grain – was working at the same horse power. From the number of stones being driven, the horse power per stone was established.

The only way Smeaton could really prove his data was to build his own mill, which he did in Leeds to crush flint. Coulomb also wanted to 'borrow' a mill for some months to properly quantify the power and efficiency, but not surprisingly, the owners declined. Even a hundred years later, in 1865, an American engineer was complaining that a similar study was still needed for both English and farm windmills to give reliable and believable data for the users, as some manufacturers were claiming far better performances than could be achieved. The true purpose, of course, for needing the power data was to design the steam engine to do the same task as the windmill – which is why the windmill owners were not so cooperative. It did not, however, stop the engineers getting their data, and the conclusion was that 3 to 4hp was needed per pair of stones. One further development was needed to the existing beam engines that could pump but not turn shafts. It was James Watt who, in 1781, invented 'sun and planet' gears to convert the up and down movement of the beam into rotation of shafts. In 1785, with his partner Mathew Boulton, they supplied the steam engines for the new Albion Flour Mill in London to drive twenty pairs of stones at 140rpm (the largest tower mill at that time would have only four pairs of stones). It was the first mill with steam power, but it would not be the last. At this time windmills were being improved, so it would be almost forty years before steam mills would be built in large numbers. In the future, steam mills, grain cleaners, flour graders, hoists and fans would all be powered by steam, as well as the stones at a steady 140rpm milling at five bushels per hour to make four bushels of flour. The installed power needed for all this was 5hp per pair of stones.

For the wind millers, the early Victorian engineers, using Smeaton's work and practical experience, produced a simple formula for calculating the horsepower developed by the sails (included in the appendices):

$$\text{Horsepower developed} = A \times V \text{ divided by } 2,000,000$$
$$A = \text{total sail area in square feet}$$
$$V = \text{wind speed in feet/sec}$$

For the typical windmill with four sails, each 30ft long and 10ft wide, and three pairs of stones, the horsepower developed at selected wind speeds is given below:

Wind speed, mph	Horsepower	Stones engaged
10	1.5	1, lightly loaded
15	5	1 or 2
20	12	3
25	24	3, fully loaded and sails reefed.

For those who wish to make their own calculations, the appendices show how these can be done. Wind-power formulae like this and the basic equations are detailed in the appendices, and examples are given on how to calculate:

Weight of air through the sail swept area
Wind force on the sails
Power in the swept area and mill efficiency
Sail area as per centage of swept area
Tip speed ratio of the sails

CHAPTER 11

The Nineteenth Century: Boom and Gloom

W indmills had ground flour from 1200 in tandem with the watermill. Both types of mill served a local population where the wheat would come no more distance than a horse and cart could make a return trip in a day, say seven or eight miles. It was ground in small batches of one sack of 280lb for a labourer, to a day's milling of thirty to forty sacks for a large farm or estate who, particularly in the early years, took the flour back to make their own bread. As towns and cities grew, merchant trade in flour for bakers established and increased, but mills continued to mill local English wheat. The stones, being a once-through process, ground the whole grain so the meal produced was coarse with fine bran and germ particles. Grain cleaning was not so good, so some tramp seeds, dust and smut also ended up in the meal. Flour dressing improved the quality of the flour, but it still remained coarse and somewhat brownish. Local English grain also had its own characteristics, having a soft chalky texture when a single grain was cut. Such grain was called weak by the miller and it yielded a flour that made a dense solid loaf. A 280lb sack of this flour would make about a hundred 4lb or quartern loaves, and the typical consumption around 1800 was 30lb of bread per week per person or 0.67 tons per year. 1 ton of bread used 0.7 tons of flour which was 85 per cent of the wheat, so 1 ton of bread was equivalent to 0.8 tons of wheat. The annual bread eaten by one person, 0.67 tons, therefore needed 0.54 tons of wheat to be grown. The population of Britain multiplied by 0.54 will give the millions of tons of wheat needed to feed them.

The mills themselves were small production units. A post mill usually had only one pair of stones which could make about one sack of flour per hour when the wind blew, which was about 25 per cent of the time or 40 hours a week, producing

5 tons or an annual output of 250 tons of flour per year per stone. Tower mills had two to three pairs of burr stones so the average windmill had two pairs of stones making 500 tons of flour a year from about 600 tons of wheat. The tons of wheat grown in any one year divided by 600 will give the number of mills in operation, at least at the early part of the nineteenth century. It should be stressed that all the statistics being used are rounded to help in the understanding of the very significant changes in milling that would take place in the next eighty years, compared to the tranquil evolution of stone milling over the last 1,800 years and windmills over 800 years. The nineteenth century would begin with protected prosperity and growth up to a climax, quickly followed by crisis and crunch.

The nineteenth century dawned with Britain still at war with France, who were blockading trade with the rest of Europe cutting off the small imports of grain so Britain had to be self-sufficient. Corn was expensive, reaching a peak of 155s per quarter (560lb) in 1812 and rarely falling below 60s. Such prices encouraged farmers to plant corn and individuals to build the average smock or tower mills at £300 to £500 to mill it. Interestingly, it was other individuals rather than banks who loaned the money, at 4 to 5 per cent interest payable half yearly with the principal sum only repaid when the mill was sold. Milling was so profitable that even larger mills were built, and the largest ever built was at Southdown at Great Yarmouth in 1812, which had six pairs of stones milling 50 tons of flour a week, an annual equivalent of about 2,500 tons of wheat milled. Mills of this size were unusual and the majority of new mills would be of the average size.

Steam-driven stone mills had been introduced by Boulton & Watt in 1788, when they built Albion Mills at Blackfriars in London. It had a 50hp steam engine to drive ten pairs, and there were plans to increase the number of stones to forty. Unfortunately, it burned down in 1791, but by the early 1800s the technology and reliability of engines had and would continue to improve so, increasingly, this type of mill would be used. Such mills were not restricted to the average of 40 hours per week reached by windmills. They would operate for 14 hours a day, six days a week, and with ten pairs of stones have an annual capacity of about 9,250 tons of wheat, but with only a few of them the average capacity of mills may have increased only slightly. With the population growing the average-sized windmills were being built in the country and urban regions, with the larger windmills and steam mills more likely in the urban areas. Data on the first thirty years of the century is given below:

Year	Population in millions	Wheat Needed (0.54 x pop, m)	Wheat Grown (statistics, mt)	No. of Mills (estimated)
1801	8.8	4.7	4.8	7,000
1810	10.5	5.7	5.5	8,000
1820	12.2	6.6	6.3	9,000
1831	13.8	7.5	?	10,000

Considering the approximations made, there is reasonable agreement between the calculated wheat needed and wheat grown data taken from *British Economic Growth 1688–1959* by Deane and Cole, but the estimate of the number of mills is probably, at best, plus or minus 10 per cent.

The war with France ended in 1815, but the Corn Laws were introduced in the same year to protect the British farmers from imports which were banned if the price of home wheat fell below 80s/qr. There was a small relaxation in 1828 when duties were set at 23s when home wheat was 64s/qr, and only 1s when home wheat was 73s/qr. The effect was complete protection of the home market and large profits for the landowners, but a high and sometimes exorbitant price of bread for the people. There was therefore pressure to not only get the price of bread down but also to remove other protective tariffs to get free trade. It was a difficult and long task, even though there were indications that home production of grain was insufficient to feed the population and mainland Britain was already dependent on supplies from Ireland. Robert Peel became Prime Minister in 1841 and began to reduce tariffs and try to repeal the Corn Laws. The disastrous failure of the Irish potato crops in 1845 required urgent supplies of grain or rice to alleviate the situation, and Sir Robert threatened to use Orders in Council to suspend the Corn Laws if Parliament did not abolish them. *The Times* of 4 December 1845 had the headline 'Doom of the Corn Laws Sealed', and the Irish potato famine and Corn Law reform was a major part of the Queen's speech on 19 January 1846. The Corn Law Bill reached its third reading, and after three days of debate was passed at 4a.m. on 16 May. Abolition would not happen until 1849, but duty was a maximum of 10s/qr at 48s/qr and a minimum of 4s/qr at a wheat price of 54s/qr, and this opened both the Irish and English markets to European supplies. However, the resentment of Parliament to the way in which Peel had forced the Bill through brought about his defeat on another Bill in June 1846 and he was forced to resign. It was the end of his political career.

The Irish potato famine continued into 1847 with considerable hardship and deprivation even with imported wheat, and, of course, there was no surplus to send to England, but this was made up by imports from Europe and even a little from America. Emigration from Ireland dramatically reduced the agricultural labour force, and even after the famine wheat from Ireland never regained its place in English supply. The landed gentry predicted the end of British agriculture with the repeal of the Corn Laws and that imports of wheat would flood in, and without tariff protection on other goods there would be grim times ahead. But the gloomy predictions did not happen! Britain began to benefit substantially from the free trade of manufactured goods, and imports of corn from Europe made up the deficit that could not be grown at home. Data to 1851 on population, wheat and mills follows:

Year	Population Millions	Wheat Needed Million Tons	No. of Mills Estimated	Wheat Imports Million Tons
1831	13.8	7.5	10,000	None
1841	15.8	8.6	11,000	0.7 (Ireland)
1845/6	Disastrous harvests in Ireland			
1846	Corn Laws repealed			
1851	17.9	9.7	12,000*	2.0 (Europe)

*This figure is considerably higher than the 10,000 often quoted as the maximum number of mills in existence at their peak in about 1880. But, the origin of this figure is *A Special Report of the Association of British and Irish Millers*, published in 1887, which has other data that qualifies the number which will be discussed in the following paragraphs.

Imports continued to make up the home shortfall, and in 1850 were about 20 per cent because the major source was Europe. This dependence on imports was emphasised in the poor harvest of 1853 but forgotten when a superb harvest followed in 1854. The Crimean War, 1854–56, prevented imports from Russia, and the Franco–Prussian War, 1870–71, interrupted supplies from France. Both wars disrupted supplies from the rest of Europe, but not from America. Fortunately, home harvests were good and there was an increasing acreage being put to wheat, which reached 8 million acres by 1874, and the yield of which was rising by improvements in agricultural practice. Yet imports, due to the needs of the growing population, had risen to 40 per cent without disturbing arable and milling affluence. The greater prosperity of the urban working population enabled them to buy a greater variety of food and become less dependent on bread. The 30lb per week at the early part of the century probably held into the 1840s and then began to fall steadily, so helping the home production of wheat meet most of the demand. (By 1924, the individual wheat needed for bread had fallen from the 0.67 tons/year of 1800 to 0.12 tons.) The growing network of railways, up from 1,857 miles in 1840 to 10,433in 1860, got the corn or flour from the countryside to the towns and cities to feed the industrial workers, whose products went by rail to the ports for export to make Britain a very prosperous place. Finally, America – and to a lesser extent Australia, Argentina, Canada and India – had yet to fully develop not only their potential for producing huge amounts of wheat, but also the means to deliver it across oceans. The American Civil War of 1861–65 delayed their investment, but the recovery after it expanded the railway system and opened up the vast expanse of the Mid West wheat-growing areas which needed export markets for both grain and flour. What imports did reach our shores continued to make up the deficit in home production caused by the increase in population, but the price of grain held at the top end of the range, 38s–53 s/qr right up to 1874, so the imports were seen as beneficial and not a threat.

Stones were the only method of milling flour up the 1850s, but technology was advancing on many fronts to exploit the greater and more reliable power of steam from coal, and the age-old milling industry was not immune from change. In Hungary, the roller-milling process emerged in the 1850s. It used five to seven pairs of steel rollers that were set at decreasing gaps so that each set did a specific part of the milling. Sieves between the roller pairs ensured that only the size that was right for the next stage progressed, and the oversize was returned to the previous pair. Being separate, the roller speed of each pair was optimised for its task so the process very effectively separated out the bran, pollards and kernel to make several grades of very fine, truly white flour. In Hungary and Eastern Europe the coarser brown flour was preferred, so they had fine white flour available for export to Britain, where the increasingly prosperous population began to develop a taste for this new product. The milling industry therefore began to use the new technology and build roller mills here in Britain. In the 1850s, they were small plants of one to five sacks per hour, but being driven by steam could work 70 to 75 hours per week and mill 500 to 2,750 tons of wheat per year – the five sacks per hour roller plant having a capacity about equal to the Southdown Windmill at 3,000 tons per year. By 1860, plants of ten sacks per hour were being built; by 1870, fifteen to twenty sacks per hour and some up to fifty sacks per hour. But this did not stop the building of windmills and large steam stone mills as the population continued to multiply.

Initially, most of the roller mills were sited in inland towns and cities, close to their markets for flour, leaving the countryside with their traditional stone ground flour from water and windmills. To meet the competition from the roller mills, windmills from the 1850s began to install steam engines to drive the stones when the wind did not blow, raising their operating times to 70 to 75 hours per week. Some used horizontal Cornish boilers with engines that could drive all the stones – a section of cogs could be removed from the brake wheel so that the turning wallower did not drive the sails when steam power was being used. Others used smaller vertical boilers and engines to drive a single pair of stones, while some added extra pairs of stones driven only and independently by steam. The older tower mills were raised in height (Market Rasen, Middle Rasen and North Leverton mills are examples) so that longer sails could be fitted for more power to drive grain cleaners, and flour graders installed in the extra floors to improve the quality, fineness and whiteness of their flour to keep their local country customers happy and themselves in business.

The stone mills, however, had a major technical problem. They had milled the soft home wheat for centuries, but it was found the harder horn-like consistency of the imported wheat could not be milled as fine as home wheat, giving a gritty texture to the flour and an even worse separation of the bran and flour. The stone mills were therefore restricted to the soft home wheat and the new roller mills dealt with imported grain which, of course, came in at the ports in the traditional 280lb sacks from Europe, though from America the 196lb cracker barrel was used. Rail transportation cost the same per sack for flour or wheat, but the wheat yielded at most 85 per cent of flour when milled, so making home-milled flour from imported wheat at a cost disadvantage to imported flour in the same town. With wheat prices averaging about 50s/qt during this period, there was a sufficient profit margin on the flour to encourage the building of an increasing number of roller plants, but that profit was progressively being squeezed.

A report of a Special Committee of the Association of British and Irish Millers of 1887 gives some interesting and important details on this period between 1851 and 1887, comparing population with men employed in the mills – whether wind, water or steam driven using stones or rollers.

Year	Population, m	Millers	Mills	Men/Mill
1851	17.9	36,076	15,000*	2.4*
1866	20	31,639	–	–
1871	22.7	29,720	–	–
1879	25.3*	24,700*	10,450	2.4
1881	26	23,462	–	–
1887	28*	21,100*	8,814	2.4

* These numbers have been calculated from the other published figures in this table.

The 1879 number for mills, at 10,450, seems to be the source of the often quoted 'more than 10,000 windmills at their peak'. Examination of the text of the report, however, shows that this figure includes 460 roller mills and that the number of millers had decreased from 36,076 in 1851 to 24,700 in 1879, indicating that the number of windmills had already fallen from a higher peak in previous years. The specific data for 1879 and 1887 gives the same average number of men in a mill to be 2.4, even though the number of windmills had declined and roller mills had increased. Roller mills employ more men than windmills but the output per man is much higher, though the change in the ratio of the different types of mill had not changed the average. In 1851, all wheat milled would be done by stones in water, post, tower or steam mills, with no roller mils. It would seem likely, therefore, that the average number of men per mill in 1851 would be at least the same, i.e. 2.4, or even a little lower than in 1879. This gives the number of mills in existence in 1851 at 15,000 – reasonable agreement with the number calculated from capacity (12,000) and lending weight to the conclusion that the peak of windmills occurred in the 1840s or early 1850s, at at least 12,000 and possibly as high as 15,000. Coincidentally, Stokhuyzen, writing about Dutch Windmills, concluded that the peak of 9,000 in Holland also occurred around 1850 with an initial slow decline that became much more rapid in the closing decades of the century, which closely follows what happened in England.

While the total number of mills was falling, new bigger windmills and roller mills were being built, so the initial slow decline must have been caused by the closure of the older post and smaller tower mills, but the total capacity for milling was increasing at least up to 1874. The 460 roller mills in operation in 1887 had a combined capability of milling 2 million tons of wheat per year, most of which would have been imported. The period from 1850 to 1874 has been described as the 'Golden Years of British Agriculture', and it would have been the same for the wind millers using stones to make brown wholemeal flour from soft English wheat. These years are summarised on the next page:

Year	Population Millions	Wheat Needed M Tons	Imports M Tons	Stone Mills	Roller Mills
1851	13.8	9.7	2	12–15,000	0
1861	20	10.5	2	11.7–14,000	100
1871	22.7	11	2.5	11.5–13,000	250
1874	23.7	11.5	2.5	11–12,000	400

The Golden Age ended abruptly in 1875 and was a big bad surprise for arable farmers and millers alike. The warning signs of impending doom (only with hindsight) had been there right from the beginning of the Golden Age, but their significance had been missed. The wheat growers had an expanding market and imports kindly just met the deficit to maintain their healthy profit. In 1850, the majority of imports came from Northern European ports in small ships still driven by sails – like the windmills– and a small amount came all the way from America. Gradually larger steam ships were introduced, reducing transport charges and making transoceanic movement of cargo a reality. The expansion of the home railway system here made possible the rapid distribution of all goods to both town and country and ports for export, but the same thing was happening across the Atlantic in America in a much bigger way. Their railroads fanned out from the east coast into the Mid West to open up vast acreages of prime wheat land. Interestingly, every station and halt had a multi-bladed windmill to pump water for the locomotives, and every farmstead had one to provide water for home and animals. They were being installed in their hundreds, as were large steam-driven roller plants to mill their newer higher yield wheat. By 1870, increasing quantities of American wheat and barrelled flour were being landed, particularly in London and Liverpool, but they were still part of the deficit so 'no worries' there.

The import of fine white flour from Eastern Europe from their roller mills in the early part of the Golden Age began to change the population's choice of bread from the dense brown wholemeal loaf to the softer, white spongy ones, and our own roller mills and more cleaning and sieving in windmills did initially meet this increasing demand. The stone mills, however, could not mill the harder foreign wheat, and so were restricted to the softer home varieties, but in the Golden Age there was enough of this to keep them in business. To add to the comfort feeling, there had not been a really bad harvest since 1853 and home wheat acreage had increased to 8 million by 1874 at yields of 30 bushels per acre (0.9 tons/acre), so why should the good times not continue?

1875 had a cold wet spring and a worse summer. In such conditions the wheat does not grow well, is prone to disease and wind and rain flatten the crop making it difficult to harvest, giving a low yield of poor-quality damp grain that goes off even more during storage. The yield in 1875 may well have been only three quarters to two thirds of 1874 – a fall from about 7 million tons to about 5 million of wheat to be milled – and there were eight more bad harvests in the next decade. Usually a bad harvest made the price of wheat rise because of the shortage, and offset the losses of millers and arable farmers alike. But that did not happen. It was the imports that increased, particularly from America. They shipped large quantities in large steam-driven ships, and the wheat could be unloaded onto the docks at a price lower than home-grown wheat. Wheat prices fell from 50s/qr in 1874 to an average of 45s/qr by 1884, 30s/qr by 1894 and 25s/qr by 1900. The Americans also exported flour from their very large roller mills into Britain, and that could reach the inland towns and undercut not only the wind millers using stones but also the new roller mills. Some of the roller mills were closing and so were many more windmills. A Royal Commission was set up in 1879 to report on the disastrous state of British agriculture, and in the same year the Association of British and Irish Millers was formed. The Royal Commission reported, in 1882, that the main cause of the problem was the imports of grain and flour from America, but since they provided cheap bread the Government took no action. Times got worse, and in 1887 the *Special Report of The Millers Association* was published, putting numbers to the closure of home mills. 10,400 mills in 1879 when there had already been four bad years, had, by 1887, become 8,814, of which 460 were roller mills, and despite closure of inland roller mills more port-based roller mills had been opened. It was the wind-powered stone mills that had suffered the most, and probably from 11,000 to 12,000 in 1875, by 1887 only some 8,000 were left. Arable farmers went out of the trade and the wheat acreage fell from the 8 million of 1875 to 2 million by 1910, compounding the loss of home-grown wheat to be milled by windmills which continued to close in large numbers.

This loss of mills had a dramatic and sudden effect on the millstone trade. The 10,000 windmills would have had about 60,000 burr stones and 20,000 Peak stones in use. When 2,000 closed, 12,000 burrs and 4,000 Peak stones were available over these eight years, and more followed at knock-down prices which killed the new stone trade very quickly. The stones being quarried and dressed in the Peak district of Derbyshire suddenly had no value, and the workers no jobs. They walked away and abandoned the few hundred partially made mill stones where they were. They remain, to this day, as a sad silent testimony to the sudden demise of an ancient trade.

Despite the dismal outlook for windmills after 1874, some stalwarts built new – or replaced old – ones, and in Lincolnshire alone five more were built. These were Kirton 1875, Heapham 1876, Sibsey 1887, Waltham 1878 and lastly, in 1889, Ludford in the high Wolds. The final windmill for flour milling by stones in Britain was built at Much Hadham in Hertfordshire in 1892, but even nature was against windmills and a gale in 1895 blew off three of the eight sails and it never milled again – perhaps a warning that the shift away from brown-stone ground flour to the white roller-milled product was unstoppable. The magnitude of the change is shown below for flour production and imports:

	Port Roller Mills Capacity, MT/Y	Inland Roller Mills Capacity, MT/Y	Imports MT/Y	Total MT/Y
1890	1.2	2.1	0.9	4.2
1906	2.5	1.7	0.7	4.9

The previous table demonstrates the shift of flour production to the ports to take advantage of bulk wheat-cargo purchases directly into very large dockside roller mills, and by 1910 there were 1,000 in operation, the largest of which had capacities of up to 250 sacks per hour (200,000 tons per year). This caused the closure of the smaller inland roller mills and was a complete disaster for the small windmills using stones. Home-wheat acreage fell from 8 million in 1875 to 2 million by 1910, and the exodus of the population to the towns and cities from the countryside took away the local village market. The combined magnitude of all these factors – but mainly the roller-mill technology – was rapidly putting the windmills out of business, closing as many as ten of them every week, and what took centuries to build was destroyed in a few decades. This is shown below:

Year	Windmills
1850	12–15,000
1875	11–12,000
1887	8,000
1890	6,500
1900	3,000
1910	2,000

For the population of 45 million in 1910, 80 per cent of the grain needed was imported and made into flour by the roller mills, and by the outbreak of the First World War there were less than 2,000 working with little money being generated for repairs to keep them going. During the war, the state of these remaining windmills was so poor with regard to trash, insects, bugs and vermin contaminating the flour, that the Government passed legislation banning the milling of flour for human consumption, leaving them only cattle feed to earn a living from. At the end of the war about 350 were left in wind, and more of these were closing. The windmill for flour had died. There was little interest in preserving them in the hard times of the 1920s, and the millers could not afford to maintain them, so more ceased working and became ruins. A very small number of people were trying to raise awareness in the value of retaining and preserving some for posterity, but it must have been a lonely and daunting task. At the end of the Second World War there were only about ten in wind and a further forty in reasonable condition. Interest in wind milling had grown considerably and it is a great compliment to the enthusiasts that there are now some 100 windmills in sail, many of which can make flour by the traditional mill stones. These are listed in the *Mills Open* book of the Society for the Protection of Ancient Buildings, so they can be easily found and visited – the remains of others have been recorded and published mainly by region.

CHAPTER 12

The American Farm Windmill

These were first used about 1840 to 1850 and were crude all-wood structures needed for pumping water in the remote areas as the West opened up to the settlers. They were like the fantail of the English windmill, with several single-angled wooden slats as sails built on open wooden towers to pump water from wells for use in homesteads, watering livestock and water for the engines on the growing rail network that brought even more settlers. They worked when the wind blew and were moved into the wind by hand. One revolution of the sail wheel made one stroke of the connecting rod from the crank at the top of the tower to the pump at the base. These wooden windmills were about 10m high on a 3m² base, with a 2.5 to 3m-diameter sail rotating at about 75rpm to pump ten gallons per minute. They were so essential to the development of the West that they did not remain simple crude structures for long.

Perhaps a new immigrant from Essex had seen young Henry Chopping's model of a multi-bladed windmill in the early 1840s and took the idea to America. This can only be conjecture, but in 1854 Daniel Halladay invented the multi-bladed windmill that reefed itself in gusts or strong winds so it could be left completely unattended. As with the development of English sails in the late eighteenth century, the early controls for automatic reefing on these multi-bladed windmills were complex and not so reliable. Simplification came in 1867 by the Revd Wheeler (he was well named) whose system turned the complete sail wheel out of the wind. It was very reliable and became the standard not only in America but around the world. By the 1880s, as the corn mills declined in England, the American windmill, also called the farm windmill or multi-bladed windmill, was being installed in their hundreds (Figure 55 and Colour 21). The gust control of Halladay and Wheeler are shown in Figure 56.

As Smeaton had evaluated the English sail in the eighteenth century, Thomas Perry in 1883 made an even more detailed study of the multi-bladed mill. He introduced curved metal blades that could be stamped out of sheet and could be mass produced at lower cost. Using more of these curved blades he increased the efficiency and power from the mill, and with reduction gearing of 2.5 to 1 to the pump, more water could be pumped from greater depths. By 1900 the all-wood had been replaced by the all-metal, and they were in use in any remote part of the world. The common size was now a 15m tower on a 3m² base with sails of 3m diameter and 30 to 40gpm could be pumped to a head of 30m by a mill working at 0.5hp or 0.4kW.

Some large machines were in use here at home, and J. Wallis Titt of Warminster had several sizes available with wheel diameters from 2 to 12m on 9 or 12m towers. The largest unit would pump 100gpm to a height of 35m in a 15 to 20mph wind working at 10hp or 7.5kW. Such a unit cost £350 in 1905. It had a fantail, as on corn windmills, to keep the wheel into the wind, and the principle of Meikle's spring-loaded shutters was used to spill wind in gusts and high winds to keep the power reasonably constant. On the largest 12m-diameter wheel were two rows of fifty blades pivoting near the leading edge. The blades, made of galvanized sheet steel, rotated at 50rpm and bevel gears (like the brake wheel and wallower of the corn mill) turned a central shaft that went down to ground level to drive a scoop wheel for drainage or an enclosed pump for water supplies.

The small units with up to 6m-diameter wheels were 'direct acting'. The rotation of the wind shaft was geared down 2.5 to 1, converted by a crank to impart up and down movement to a central rod that was linked to a lift pump for drainage or a pressure pump for supplies. The smaller wind pumps did not have pivoted and spring-loaded sails. They were fixed rigidly onto the two circular steel bands that formed the wheel. It was the canting wheel that dealt with increase in wind speed and gusts (Figure 57). The sails were kept into the wind by a tail vane which was offset from the axis of the sail wheel and not fixed at right angles to it but can pivot to let the wind wheel change its angle to the wind. It was normally held at right angles by a helical spring on the vertical pivot which kept the wind wheel into the wind in steady moderate winds. As the wind strengthened, the force of the wind overcame the holding force of the spring and turned the wind wheel out of the wind and at an angle to it, so taking less energy from it. When the wind eased, the spring on the pivot returned the sail wheel back to fully face the wind. For both sizes of windmill, these reliable wind-control systems meant that they could be left unattended for long periods of time. They also needed very little maintenance and had long lives – some lasted seventy to eighty years.

The range of sizes is given on the following page:

Sail wheel diameter m	Water raised 35m in 10mph wind gph	Speed of sail wheel rpm	Power in 15/20mph wind hp	kW
2	80	125	0.05	0.04
3	200	110	0.12	0.09
5	600	80	0.6	0.45
7.5	2000	60	1.35	1.0
12	6000	50	2.5	1.9

Corn windmill 12		8/15	12/15	9/12

Even the largest farm windmill was much less powerful than the traditional English windmill for corn or its equivalent for pumping water. Here, high flows up a few metres for drainage was usual, whereas in America and other parts of the world, lower flows from deep wells was more often needed. As the American West opened up, more and more of these multi-bladed farm windmills were installed. Every farmstead had one and every rail stop needed one for the locomotives that hauled the trains that brought even more settlers.

By the beginning of the First World War there would be nearly a million of these windmills in use, and by its end the returning heroes wanted more than water on the farm. They wanted electric light like their city cousins. Industrialisation had increased apace in the war, and with it knowledge and experience of machines that would normally have taken a generation or two to get out into the farming areas did so in a few years. Those returning from the war had this knowledge and experience and so they were able to use dynamos from lorries or cars and fix them to the windmill, and by belt drives charge lead acid batteries to get 12v DC for lighting. This would have been a vast improvement on candles or pressure paraffin lamps. There were major difficulties to be overcome. A vehicle engine runs at about 1,500rpm to drive the dynamo, but the farm windmill turns at no more than 100rpm, and gearing up to get the speed would reduce the already small amount of power even more, so only a few hundred watts of power were available. The provision of water pumping was the driving force for tens of thousands more farm windmills to be installed, along with a frustratingly small amount of power for the farmer.

Marcellus Jacobs was one such farmer in Montana and he wanted more electric power from his windmill so, in the early 1920s, he began trying modifications to get round the limitations imposed by traditional construction. He experimented for several years, and by 1927 he had designed and built a windmill on a 15m-high tower with sails of 4.5m diameter. His neither had sails like the multi-bladed farm windmill nor the traditional English windmill. It had three slender blades like – but much larger than – the wooden aeroplane propellers in use on the biplanes of the time, which Marcellus had learned to fly (Figure 9). He must have been inspired by these propellers and impressed by the power they generated to make the plane fly.

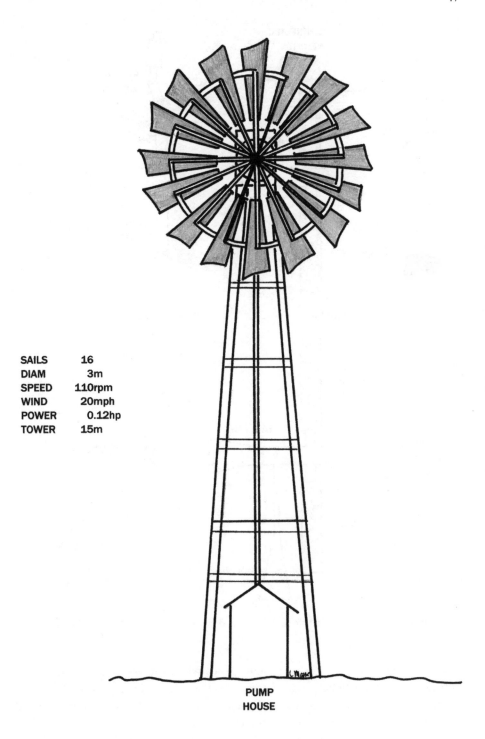

SAILS	16
DIAM	3m
SPEED	110rpm
WIND	20mph
POWER	0.12hp
TOWER	15m

**PUMP
HOUSE**

Figure 55. American farm windmill.

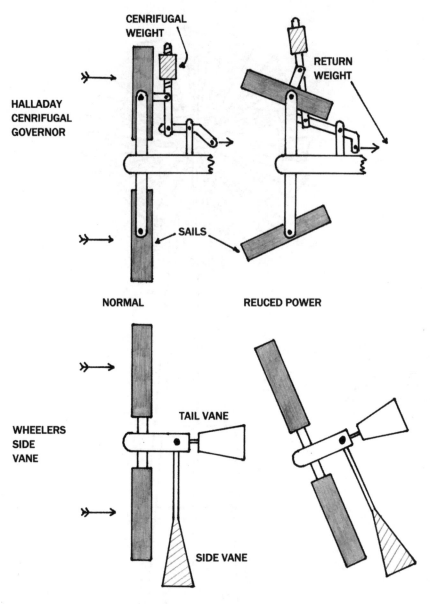

Figure 56. American farm windmill, speed and gust control – Halladay and Wheeler.

Even so, he could not get very high speeds when the wind turned his blades instead of an engine, so he designed a dynamo that operated at low speed but could produce 110v DC to charge a 21kW capacity battery pack that would meet the needs of most farmsteads. The complete unit was priced at $1,000 and could be installed in two or three days. The Jacobs' system, and others similar to it, could now be generating 2 to 6kW from 4 to 6m–diameter sails. Being DC, however, they could

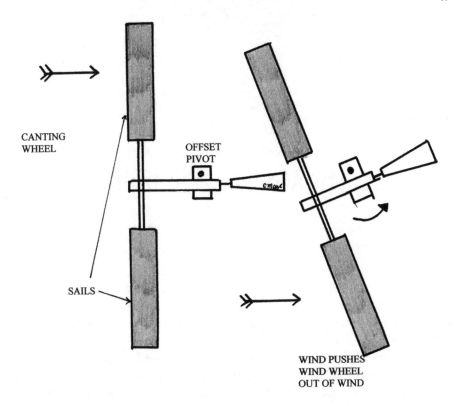

CANTING
WHEEL

OFFSET
PIVOT

SAILS

WIND PUSHES
WIND WHEEL
OUT OF WIND

Figure 57. English Farm Windmill, speed and gust control – canting wheel.

not power the growing number of mod cons that were being introduced for high-voltage AC provided by the large utility companies. Such items remained a dream for the rural household, no matter how large and wealthy it was.

Water pumping plus lighting remained the main use of the multi-bladed farm windmill, and at their peak, in about 1930, some 6 million units were in use in America with more in South America, Australia and India and any other remote place. The slender three-sailed windmill took on the roll of small-scale power generation for remote farmsteads, but rural electrification was beginning, and whenever the utilities-distribution system reached an area, both it and the multi-bladed types of windmill were redundant as the small low-voltage DC system was no match for 250v AC. The investment and manufacturing capacity needed to complete the job and the depression of the 1930s meant that there would be a place for the multi-bladed and slender bladed windmill for some time to come.

Even now, in the twenty-first century, a few multi-bladed windmills still operate for water pumping, but they are a rarer sight in England than restored corn windmills. In remote areas, both types continued to find a limited use but Jacobs' revolutionary use of the propeller-like sail would lead, half a century later, to a fresh large-scale expansion in the use of wind power.

CHAPTER 13

The Modern Windmill

These are usually referred to as wind turbines since they are generally large and generate AC electric power for a grid, as do coal and nuclear stations. Their size actually ranges from a few kilowatts to several megawatts and it is the latter, because of their striking appearance, which are well known (Colour 22, 23 and 24). They are still windmills in the generic sense, as anything with sails that mills, pumps, saws or generates power, i.e. turns machinery for whatever purpose, is a windmill. While these turbines are regarded as modern, their origin is more than a century old, as it was Moses G. Farmer in 1860 who patented a windmill with a dynamo to power a few lights.

During the First World War, the aeroplane invented by the Wright brothers less than a decade before became a vital factor in the battles, and to make it fly faster and higher than the enemy could the science of aerodynamics was established. After the war, this science of wing and propeller design was applied to peacetime uses such as the windmill, and some very novel concepts were put forward. Some only got to the model stage but one or two were built to get free power from the wind. The traditional English windmill had virtually died and the American farm windmill had serious limitations on size and power. It was extremely useful for pumping water and generating a few kilowatts of DC power for lighting as well as the new-fangled things called wireless sets, and a useful test bed for the practical inventor who also wanted the free power from the wind.

Kumme in Germany in 1920, using the principles of aeroplane wing design, built a six-sailed windmill for power production (Figure 8). The narrow sails of 20m diameter were parallel sided with sections just like wings, and the power was transmitted to the ground by a vertical shaft directly into a vertical generator. The braking effect of the generator was big enough to turn the sails out of the wind, overcoming the small windmills on the cap that were supposed to keep them into the wind. The power produced was, therefore, much less than was required for viability, so the project was abandoned.

The frontier breakers did not confine themselves to traditional windmill lookalikes. Magnus had discovered, in 1850, that if a cylinder was rotated in a wind, a sideways thrust was generated (Figure 11). In 1925, Fletner decided to use this principle as a prime mover. He designed and built a ship called the *Baden Baden* that had two vertical, slim, rotating cylinders at each end of the deck. To prove his point, he sailed the ship across the Atlantic to New York. He followed this up in 1926 with a land-based windmill with four sails of 20m diameter on a 32m-high tower. Each sail was a slightly tapering cylinder 5m long, averaging 0.75m diameter, that was rotated by an electric motor in the hub. The thrust on each cylindrical sail turned the sail wheel, which powered a dynamo whose DC electric not only provided power for the sail motors but also charged a battery array. It was rated at 30kW in a 23mph wind, but did not reach this target. The engineering of producing and maintaining simultaneously spinning and rotating sails is 'mind bending', and no more of these were made (Figure 8).

Madaras also used the Magnus effect. On a circular rail track he proposed a fleet of flat bed trucks, each with a driven cylinder 30m high and 8m in diameter spun by an electric motor, and the thrust generated pushed the cars along the track. On each of the car axles was a generator which sent power to a pick-up rail and then to a control/distribution centre. The cylinders had to reverse rotate twice on each circuit of the track, which would have needed a large flat area so that the radius of the track was big enough not to impose a large frictional drag on the car wheels that would have significantly reduced the efficiency of generation. The electrical resistance between the pick-up rail and its contactor would also have been large, reducing still further the efficiency. Reversing of the cylinders also needed high energy that would be taken from the generators. Despite these perceived drawbacks, Madaras got sufficient backing by October 1933 to build a single pilot cylinder on a concrete base at Burlington, New Jersey, to quantify the thrust generated. It attracted a large audience, but the thrust measured did not match the size of the crowd and mobile windmills for power never progressed.

Another variation on the vertical-axis windmill was the Savonius invention (Figure 11). In this, two half cylinders like long troughs are mounted on a vertical axis with the two troughs separated by half their diameter so that the wind is made to follow an S-shaped path between them. The jet of the accelerated exit wind causes the sail to rotate to turn a generator. It was calculated that it would be about 30 per cent efficient, but a cylindrical sail 90m high and 9m diameter would be needed to generate 1,000kW in a 30mph wind (this assumes that the cylinder would remain upright under these wind conditions).

The Russians also tried their hand at generating power, and in 1931, after surveying the wind behaviour and deciding on a site at Yalta on the Black Sea, a 100kW windmill was built (Figure 8). It had four sails based on aeroplane wing design with a diameter of 30m on a tower of the same height. Like conventional windmills, the sails were on the up-wind side of the tower and the thrust from the wind was opposed by a long gantry from buck to ground, rather like the stairs on a post mill. At the foot of the gantry was a car that ran on a circular track. The position of a vane on the buck controlled electric motors on the car to keep the mill facing into the wind. It had a mechanical sail-speed control operated by a governor to increase pitch (and hence drag) in increasing winds, and counter weights to put the pitch back to optimum (approximately 15 degrees) as the wind eased. Control was no better than plus or minus 15 per cent, but it did generate power and fed a local grid with a peat-burning power station already on it. Plans to add more and bigger units were never completed.

The most imaginative proposal was by Honnek of Germany in 1933 (Figure 10). This was a massive 50 megawatt unit standing on a 300m-high tower with five 75m-diameter windmills attached to a rotatable cap. The total height would have been 425m! The tower design was similar to that of the Zeppelin air ships to give a high degree of stiffness but with some flexibility and resilience to withstand the buffeting of the wind. It had to be strong, as the rotating mass was 150 tonnes with a span of 225m. The estimated cost was £150,000, but it would have made power at a farthing a unit (a farthing was a ¼ of an old penny so in today's money, 1kW would have cost 0.001p), so the incentive to build one was great. Zeppelins were built and flown around Europe and across the Atlantic, but Honnek's windmill never got beyond the model stage. This was a very wise decision as the disaster of the Zeppelin *Hindenberg* in New York in 1937 may well have been the fate of a full-sized Honnek windmill in a similar storm.

In 1929, a French electric company, built at Le Bourget, a mill designed on the Darrieus principle (Figure 8). The tower was 30m high and the two sails were 20m diameter and tapered. They were also down-wind of the tower and inclined backwards so that the swept area formed a shallow angled cone. Such a design aligns the sails into the wind without the need for a luffing drive and reduces the bending stress at the hub/sail joint. Darrieus also patented vertical-axis windmills with four sails that were attached only at the top and bottom of the axis. They were curved or diamond-shaped and hence looked like an egg or double cone when rotating. Like his down-wind sails, these vertical-axis windmills needed no wind alignment as they worked with the wind from any direction. The vertical shaft drove a vertical generator without the drawback of holding the sails out of the wind like the Kumme invention did, but neither lead to further commercial units.

In all these attempts to get power from the wind, no one really new how much power could be got from it. In 1928, Albert Betz, a German aerodynamicist, studied the theory of wind behaviour and calculated that the maximum energy that could be obtained was 16/27 or 59 per cent of the total energy passing through the swept area of the sails. This occurs when the sails reduce the velocity of the wind by 1/3 in extracting the energy and whatever the design of the sail, 59 per cent is the limit of power extraction.

While the inventive scientists and designers of Europe were enjoying themselves with strange concepts, as mentioned in the previous chapter, a down-to-earth American farmer, named Marcellus Jacobs of Montana, struggled in the 1920s with his multi-bladed farm windmill to get more power from it to run his wireless set and a few light bulbs. From 1920, and by 1927, he came up with a series of improvements for small farm-based windmills:

1   A revolutionary sail design that had three slender sails like a biplane propeller!

2   A tail vane to hold the sails into the wind.

3   Speed control by pitch variation operated by centrifugal weights and return springs that changed the pitch of all three sails the same amount at the same, time to maintain stability of the wind wheel at any setting.

4   The use of a longer dynamo with more coils to give 110v DC at a relatively low speed. This charged a 21kW-capacity battery pack that was more than adequate for farmstead use.

These, and similar types, were mass produced in their tens of thousands and sold for only $1000, and could be erected in a couple of days. They were not big enough to generate 250v AC and so could not power the growing range of domestic and industrial equipment suitable for homes and farms that were connected to the grid.

Jacobs, however, had made a very significant revolution in sail design equal to the achievements of Smeaton, Meikle, Cubitt and Perry. The slim three-bladed windmill making a few kilowatts in the 1930s would, half a century later, become the basis for the megawatt wind turbines of the 1980s. Their use in the American West was very widespread because rural extension of the grids needed not only equipment but also significant investment which, during the years of the 1920s and 1930s depression, was very hard to get. It has been estimated that, at their peak in the 1930s, there were some 6 million of these DC dynamos in use. The big 250v AC generators were still an unattainable dream.

It needed another American, despite the failures of the European designers and inventors and the limitation on size of the successful DC Jacobs at home, to decide to consider 250v AC for wind power at his home. His name was Palmer Coslett Putnam. In 1934 he wanted 250v AC for his new country house in Cape Cod. Even with his wealth, the dollars quoted by the nearest power company to extend the grid and the cost of power from it, he regarded as excessive, and so decided to look at windmills to provide his power. He was a graduate of MIT (Massachusetts Institute of Technology) and so well able to understand and evaluate the electrical and mechanical aspects of wind-power generation. From the long list of appliances, lights and tools that needed 250v AC power, he quickly established that the size of the generator to meet the peak load when all these were working was not only large and hence expensive, but that he would still have to have – and pay for – grid power when the wind did not blow because AC power could not be stored. At this point, most men would have given up, but not Palmer Putnam! He began an in-depth study of the machines that had been tried for wind-power generators and how wind generators should fit into an overall power-supply scheme.

The basic condition to be met was that houses and industry needed power 24 hours a day, 365 days per year, and at the flick of a switch. This could not be done by wind alone, so a windmill generator had to be linked with a grid system already fed by power from coal or water, which had been tried in Europe but had not been adopted. In such a system, the windmill is not restricted to 250v AC peak load of a house or single user. It could be designed to be of a size to produce power economically and send that power into a high-voltage grid system for any customer. The windmill could therefore be big, but must be mechanically and electrically reliable, and be located at a place where the wind blew regularly in the required speed range. With this as his design remit, he began what would be a three-year study of the recent ideas, models and plants that had been tried. He also obtained what wind data there was over the Vermont region of America, and by 1937 Palmer had designed a windmill generator for high-voltage grid power. It had a capacity of 1.25MW, the characteristics of which are given below.

PUTNAM'S WINDMILL SPECIFICATION

Capacity	1.25MW
Sail diameter	53m
Sail length	20m
Sail width	3.5m
Sail section	Aerofoil
Sail shape	Rectangular
Sail weight (each)	8 tonnes
Sail material	Stainless steel sheet on carbon steel ribs
No. of sails	2
Location	Downwind of tower*
Design wind speed	17mph
Design shaft speed	29rpm
Angle of wind shaft	15 degrees
Gust protection	Variable Coning *
Luffing	Motor-driven activated from pilot vane on cap
Speed control shaft	Pitch control of sails activated from speed of wind
Tower height	36m
Tower weight	70 tonnes
Weight of cap, sails, etc.	210 tonnes
Curb diameter	3m

* Variable coning of blades down-wind of the tower was initially conceived by Darrieus and tried by the French in 1929, so it is likely that the results of those trials helped Putnam

to decide that this was the type of windmill that was most likely to succeed. The sails were hinged at the hub and had a normal cone angle of 5 degrees out of the wind. This coned attitude helps to keep the sails into the wind, and increasing winds increase the cone angle to a maximum of 15 degrees, which not only reduces the effective diameter of the sails but also tilts them out if the wind, thus taking less power from the wind.

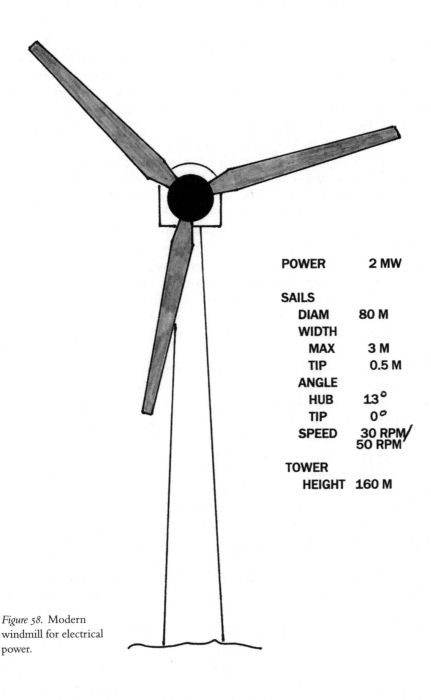

POWER	2 MW
SAILS	
DIAM	80 M
WIDTH	
MAX	3 M
TIP	0.5 M
ANGLE	
HUB	13°
TIP	0°
SPEED	30 RPM / 50 RPM
TOWER	
HEIGHT	160 M

Figure 58. Modern windmill for electrical power.

The specification on the previous page is a brief summary of what was a complete engineering design and build specification, with costing estimates that could be presented to companies to consider. Again, with the help of his contacts he sought backers to build his windmill and evaluate its performance in an integrated power-supply system, and in 1939 the Morgan Smith Co. decided to build it. They had the cooperation of The General Electric Co., Massachusetts Institute of Technology and the Central Vermont Public Services Co. (the hydro-electric company in that area). A site at 600m elevation near Rutland in Vermont, called Grandpa's Knob, was selected and construction began. Despite the outbreak of the Second World War, the Smith–Putnam windmill was completed by August 1941, and on 19 October began supplying power to the grid. It clocked up a few hundred hours of operation, but in February 1943 one of the main bearings failed, and with wartime restrictions it took two years to replace it. During that time, the sails were locked in the vertical position through two winters of buffeting high winds with ice and snow.

It began operation again in early March 1945, but on 26 March what had been the upper sail during shutdown failed at the hub connection and was hurled 100m or so across the hillside. The investigation showed that stress corrosion cracking across 90 per cent of the sail root during its enforced shutdown was the prime cause, followed by fatigue cracking (a metallurgical term for repeated bending) which completed the failure. The actual electrical performance was good, at an efficiency of 30 per cent, but the investment costs were high at $190 per kilowatt compared with $125 for coal or hydro power. It had cost the private company of Morgan Smith over $1 million to fund the evaluation, and it would have required a similar amount to develop ways of reducing the investment costs at a time when coal and oil for power were cheap and abundant. They therefore decided it would not be worthwhile and abandoned the project.

In the meantime, electrification was sweeping across the farmlands of America and, when it arrived, both the multi-bladed and the newer two or three slim-sailed windmill were redundant. By 1950, of the 6 million in existence in the 1930s, very few remained. The market for them had moved to fresh remote areas in Australia, India and South America, where mass-produced kits were supplied for the user to build himself until grid power reached him.

An exciting new source of energy had also become available. The enormous and devastating power of the atomic bomb was being investigated as a controllable peacetime provider of nuclear processes to make electricity. The reactors would last a very long time and their power would be cheaper than from coal or oil – or so it was thought. Large government funding was used to develop the physics and engineering needed to get to large commercial stations. Coal and oil remained cheap and the capacity of power stations reached 2,000MW with individual generators of 500MW, so who wanted small windmills for power that could only operate for about 25 per cent of the time when the wind blew in the wanted speed range.

By the 1960s, many nuclear power stations were operating but were not proving to be quite as cheap a source of electricity as hoped, and the problem of waste treatment

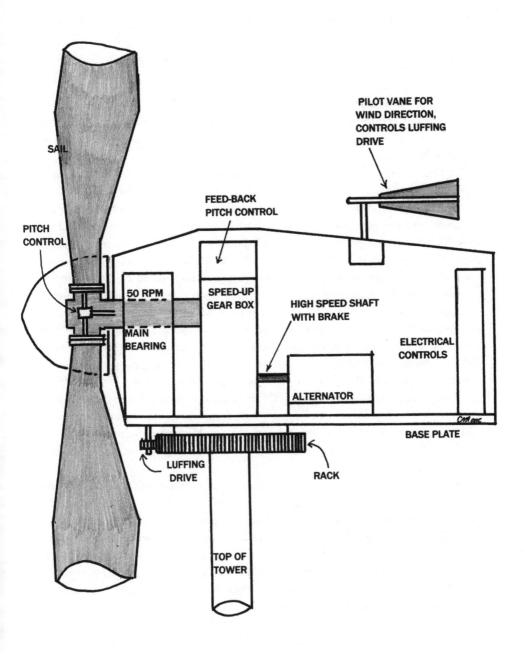

SAIL

PITCH
CONTROL

FEED-BACK
PITCH CONTROL

PILOT VANE FOR
WIND DIRECTION,
CONTROLS LUFFING
DRIVE

50 RPM

SPEED-UP
GEAR BOX

HIGH SPEED SHAFT
WITH BRAKE

MAIN
BEARING

ELECTRICAL
CONTROLS

ALTERNATOR

BASE PLATE

LUFFING
DRIVE

RACK

TOP OF
TOWER

Figure 59. Modern windmill, nacelle.

was beginning to be realised and quantified. By the early 1970s, energy dependency was:

Coal – 25 per cent   Oil – 45 per cent   Natural Gas – 20 per cent   Nuclear – 10 per cent

But the oil crisis of 1974 made the 45 per cent dependency on petroleum products politically and economically unacceptable, so other renewable sources of home-based energy were needed and wind power was a prime candidate. The windmill would be born again.

The basic science of wind power was well known, and the experience from small windmills and the larger Smith–Putnam windmill were available. New materials and manufacturing techniques proven in other industries only needed translating into windmill power generation, and it did not take long. Several variations of windmills were tried, but soon the three narrow-sailed new post windmill dominated and the Western world began to install them in significant numbers from 1980. They are now very sophisticated remote-operated mini-power stations of up to 5MW capacity (Figure 58). With computer management (Figure 59) they can be aligned into the wind, but the pitch control of the sails turns them parallel to the wind to keep them still until the wind speed reaches the designed 'cut in' speed. The pitch control then angles the sails and they begin to spin and accelerate to the operating speed of about 50rpm. A speed-up gearbox in the nacelle (the buck of the original post mill) drives a high-speed shaft to the alternator at a speed appropriate for the grid conditions (i.e. voltage and frequency of the AC). Other electrical controls then synchronise the alternator's AC wave form with that of the grid, and when matched the alternator's power locks into the grid at low output. The pitch control then moves the sails so that more energy is taken from the wind and alternator output increases to design level.

As wind speed and direction vary, the sail pitch and nacelle luffing controls automatically compensate and keep the alternator revolving at its design speed. If the wind speed increases too much and reaches shutdown speed, the pitch system controllably turns the sails out of the wind and with the help of additional braking slows them down, the computer takes the alternator off-line and the windmill comes to a controlled stop with the sails in the park position, i.e. parallel to the wind. Shutdown will be initiated for any fault condition, or by grid control if the power is not required. The modern windmill for power operates remotely and automatically with absolute minimum of human intervention. With materials such as fibre glass, GRP (glass reinforced plastics) and carbon fibre in use for the sails, maintenance is also kept to a minimum, as, being non-metallic, they do not suffer from corrosion and fatigue.

In making power, the modern windmill produces neither ash, carbon dioxide, acid rain, warm water from coolers nor nuclear waste. Its source of energy – the wind – is inexhaustible and immune from being cut-off by political or industrial conflict. Traditional windmills would only operate for about 25 per cent of the time, but their modern equivalents will work for double that time. Even so, a single turbine is producing electricity for only half the time. However, over a wide enough area or country, the wind will be suitable somewhere all of the time. So, with a number of turbines on dispersed and selected windy sites, continuity of supply for about one third of the total wind capacity can be maintained, but coal and nuclear stations are still vital for the majority of supply because electricity has to be used at the moment of its production. Small variation in demand is met by turn-up or down of individual units, and large changes by bring-on or take-off of several units. Buffer storage of power is not possible in the megawatt range – but who knows what tomorrow will bring?

CHAPTER 14

How the Windmills Compare

In this section the use, basic data and performance of the three types of windmill will be used to highlight the similarities and differences – the most striking of which is the sails (Figure 60).

Basic Data Use	English Flour	Farm Water	Modern Power
Power: HP	12 to 16	0.1 to 6	675 to 6,750
kW	9 to 12	0.08 to 4.5	500 to 5,000
Sails: No.	4 to 8	16 to 100	2 or 3
Shape	rectangular broad	rectangular slender	tapered slender
Efficiency: %	18 to 25	10 to 15	25 to 30
Sail speed: rpm	12 to 18	25 to 110	30 to 50
Wind speed range mph	8 to 40	8 to 40	15 to 50
Tip Speed Ratio, TSR	1 to 3	1.5 to 2.5	10 to 13

These figures should be taken as a reasonable guide, but because of the individual nature of even modern windmills, do not be surprised if you collect data on an

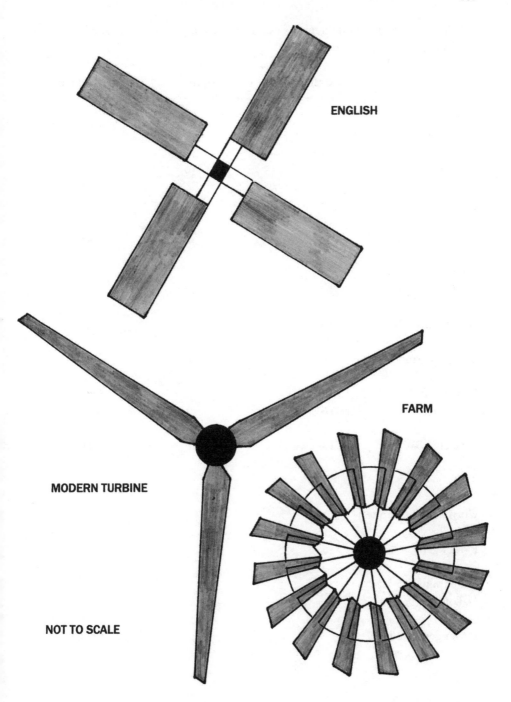

Figure 60. Sails of the windmills.

individual mill that the odd item lies outside these ranges. Specific examples of three typical mills are given below.

	English	**Farm**	**Modern**
Sails. No.	4	80	3
Length, m	8	2.5	38
Width, m	2.5	0.15	2.25
Area, m²	80	30	250
Arm length, m	9	7.5	40
Swept circle area, m²	255	45	5,000
Sail to swept area ratio %	30	65	5
Best wind, mph	20	20	30
Sail speed, rpm	15	30	40
Power: HP	12	1.3	2,700
kW	16	1.7	2,000
Efficiency %	18	12	26
TSR	2.2	1.8	12.2
Corn milled (2 stones)			
Sacks/hr	2	–	–
Flour kg/hr	210	–	–
Water pumped			
Gallons/hr	–	12,000	–
Power generated kW	–	–	2,000

APPENDIX I

Wind Power Formulae

The following letters denote certain parameters that are needed in calculating the power in the wind:

P = power in the wind
m = mass of air
v = velocity of air
e = density of air
t = time
d = sail diameter
A = sail area or swept area
π = pi (3.142), the ratio of the circumference of a circle to its diameter.

BASIC EQUATIONS

$$\mathbf{1} \quad A = \frac{\pi}{4}\,d^2 \qquad\qquad \mathbf{2} \quad m = A.e.t.v \qquad\qquad \mathbf{3} \quad P = \frac{1}{2}\,.m.v^2$$

$$= \frac{\pi}{4}\,d^2.e.t.v$$

Put equations 1 and 2 into 3

$$P = \frac{1}{2}.\left[\frac{\pi}{4}\,d^2.e.t.v\right]v^2 \qquad\qquad P = \frac{1}{8}\left[\pi e.t.d^2\right]v^3$$

The quantities in brackets are constants, so P is proportional to v^3
Or, the power in the wind is proportional to the velocity of the wind cubed
In 1759, John Smeaton came to this very same conclusion!

Calculations of the Weight of Air Passing Through the Swept Area of the Sails

All numbers are rounded to the nearest significant number.

Wind speed is 15mph or 6.5 m/sec (v); sail diameter is 18m (d)

$$\text{Swept area, A,} = \frac{\pi}{4} \, d^2 = \frac{3.142}{4} \times 18 \times 18$$

$$= 255 \text{ m}^2$$

$$\text{Volume of air} = \text{area (A)} \times \text{wind speed (v)}$$

$$= 255 \times 6.5$$

$$= 1654 \text{ m}^3/\text{sec}$$

$$\text{Density of air (e)} = 1.225 \text{kg/m}^3$$

$$\text{Weight of air} = \text{volume} \times \text{density}$$

$$= \frac{1654\text{m} \times 1.225 \text{ kg/m}}{1000\text{kg}}$$

$$= 2 \text{ tonnes per second}$$

Calculation of Wind Force on the Sails

Measured wind pressures at selected speeds:

Wind speed Mph	Resulting pressure lb/ft²	Kg/m²
12	0.75	3.7
25	3.13	15.5
50	12.5	62

Windmill data: arm diameter – 18m; sail length – 8m; sail width – 2m; no. of sails – 4.

$$\text{Total sail area} = 8 \times 2.5 \times 4$$
$$= 80\text{m}^2$$

Total force on sails = area $\times$ pressure

Wind speed mph	Total force on sails tonnes
12	0.3
25	1.2
50	5

Power in the Swept Area of the Sails and the Efficiency of the Mill

Watt's standard horsepower

$$1hp = 33,000ft\ lb/min$$
$$= 750W$$
$$or\ 1kW = 1.35hp$$

Power (kW) passing through the swept area of the sails can be calculated by:

$$P = \frac{0.005.A.v^3}{1,000}$$

Where A is the swept area in square feet and v is the wind speed in mph.

EXAMPLE 1   English windmill with 60ft-diameter sails working in a 15mph wind with four pairs of stones working at 3 to 4hp each.

$$Swept\ area\ (A) = \frac{\pi}{4}\ d^2\quad = \frac{3.142}{4} \times 60 \times 60$$

$$= 2,828ft^2$$

$$P\ (kW) = \frac{0.005 \times 2828 \times 15 \times 15 \times 15}{1000}$$

$$= 48kW$$
$$or = 65hp$$

$$Efficiency\ \% = \frac{power\ used \times 100}{power\ available}$$

$$= \frac{12\ or\ 16}{65} \times 100$$

$$= 18\ to\ 24\%$$

EXAMPLE 2 American farm windmill with eighty sails, 25ft diameter, working in a wind of 15mph with sails rotating at 60rpm generating 1.3hp for pumping water.

$$\text{Swept area (A)} = \frac{\pi}{4}\,25 \times 25$$

$$= 490\text{ft}$$

$$\text{Power (kW)} = \frac{0.005 \times A \times v}{1,000}$$

$$= \frac{0.005 \times 490 \times 15 \times 15 \times 15}{1,000}$$

$$= 8.3\text{kW or } 11.2\text{hp}$$

$$\text{Efficiency} = \frac{1.3}{11.2} \times 100$$

$$= 12\%$$

EXAMPLE 3 Modern power windmill with three sails, 80m diameter (265ft), sail speed 40rpm, wind speed 30mph and output of 2,000 kW.

$$\text{Swept area} = \frac{\pi}{4}\,265 \times 265$$

$$= 55,150\text{ft}^2$$

$$\text{Power (kW)} = \frac{0.005 \times 55,150 \times 30 \times 30 \times 30}{1,000}$$

$$= 7,500\text{kW}$$

$$\text{Efficiency} = \frac{2,000}{7,500} \times 100$$

$$= 27\%$$

Sail Area as a Percentage of Swept Area

EXAMPLE 1   English windmill with four sails 27ft long and 7ft wide on arms 30ft long (60ft diameter):

$$\text{Sail area} = 27 \times 7 \times 4$$
$$= 756\text{ft}^2$$

From example 3.1   Swept area $= 2,828\text{ft}^2$

$$\text{Sail area as percentage of swept area} = \frac{756}{2,828} \times 100$$

$$= 27\%$$

EXAMPLE 2   American Farm windmill with eighty sails, 8ft long and 6in wide with a wind-wheel diameter of 25ft:

$$\text{Sail area} = 8 \times 0.5 \times 80$$
$$= 320\text{ft}^2$$

From example 3.2   Swept area $= 490\text{ft}^2$

$$\text{Sail to swept area \%} = \frac{320}{490} \times 100$$

$$= 65\%$$

EXAMPLE 3   Modern power windmill with three sails, 80m diameter (265ft), sail length 125ft, average width 7.5ft

$$\text{Sail area} = 125 \times 7.5 \times 3$$
$$= 2,810\text{ft}^2$$

From example 3.3   Swept area $= 55,150\text{ft}^2$

$$\text{Sail to swept area \%} = \frac{2,810}{55,150} \times 100$$

$$= 5\%$$

Tip Speed Ratio – TSR

This is the current method of comparing the performance of modern tapered-sail windmills rather than the overall efficiency, but it is used here for all three types. It is the ratio of the sail tip speed to the wind speed.

EXAMPLE 1   English windmill with sail diameter of 60ft in a wind speed of 15mph (22ft/sec) turning the sails at 15rpm:

$$\text{Circumference of sails} = \pi d$$
$$= 3.142 \times 60$$
$$= 188\text{ft}$$

$$\text{Distance travelled by tip in 1 second} = 188\text{ft}\, \frac{15}{60} \times \text{rpm, ft}$$

$$= 47\text{ft}$$

$$\text{TSR} = \frac{47}{22}$$

$$= 2.2$$

EXAMPLE 2   American farm windmill with wind wheel diameter of 25ft turning at 30rpm in a 15mph wind:

$$\text{Circumference of sails} = \pi d$$
$$= 3.142 \times 25$$
$$= 79\text{ft}$$

$$\text{Distance travelled by tip} = 79 \times \frac{30}{60} \text{ rpm, ft in 1 sec}$$

$$= 40\text{ft}$$

$$\text{TSR} = \frac{40}{22}$$

$$= 1.8$$

EXAMPLE 3   Modern power windmill, three sails, 80m (265ft) diameter, turning at 40rpm in a wind speed of 30mph (45ft/sec):

$$\text{Circumference of sails} = 3.142 \times 265$$
$$= 830\text{ft}$$

$$\text{Distance travelled by tip} = 830 \times \frac{40}{60} \text{ ft}$$

$$= 550\text{ft}$$
$$\text{Wind speed} = 45\text{ft / sec}$$

$$\text{TSR} = \frac{550}{45}$$

$$= 12.2$$

Mill Terminology

These are the words used to describe items in the mill. Some are used over the whole country, but sometimes different words are used for the same thing on a regional basis, and some may apply to a few mills only. The centuries have also changed the words, but each one is equally important and needs to be recorded.

AIR BRAKES Longitudinal shutters on the leading edge of some sails that reduce sail speed in strong winds.

ALARM BELL The small bell activated when the hopper runs out of grain. Colour 43.

ANNULAR SAILS Circular sail of radial shutters used on farm windmills. Figures 55, 60 and Colour 21.

ARM DIVIDE The ratio of the sail widths on either side of whip.

BACK EDGE Black, cutting or sharp edge. The near-vertical edge of the furrows.

BACK STAYS Strengthening bars across the back of sails.

BACKWINDED (tail winded). The very dangerous situation when the wind blows from behind the mill.

BALANCE WEIGHTS The pieces of lead or iron that were fixed in holes by plaster of Paris in the runner stone that ensure perfect balance. Later, the metal washers on screws in vertical or horizontal holes in the runner stone to reach the same balance. For sails, the iron plates or bars fixed to the end of the sail whip to balance the rotating sails.

BALCONY (gallery, stage.) Platform fixed to a tower mill of more than four floors so that sails can be reached for reefing. Colour 15.

BATTER The slope of the walls of a tower or smock mill. Colour 14.

BAY The section between the sails bars that hold two or three shutters. Colour 29.

BED STONE (nether, ligger.) The lower, fixed, stationary mill stone. Colour 40.

BEARING BOX (glut box.) A square box with screws on each face that can adjust the alignment of a shaft. Colour 45.

BELL ALARM See alarm bell.

BELL CRANK (crank, triangle.) The lever linking the spider to the sail bar that operates the shutters on a patent sail. Colour 33.

BERRY A grain of wheat. Figure 52.

BILL The special wedge-shaped chisel used to cut the furrows and stitches on the stones. Figure 46 and Colour 42.

BIN The storage box for the received grain, usually on the top floor of the mill.

BIRDS MOUTH JOINT A special joint used to locate the quarter bars on the cross trees of a post mill. Figure 13.

BLUE STONE Cullin stone. Millstones originating from Cologne in Germany.

BODY (buck.) The main part of a post mill. Figure 5 and Colour 1 and 2.

BOLTER (boulter, dresser, grader.) A rotating machine with sieves to remove the bran and trash from the meal and separate the flour into two or more grades of flour depending on fineness.

BRAKE (gripe.) The iron band or string of wooden shoes that encircle the brake wheel.

BRAKE WHEEL (head wheel.) The large gear wheel on the wind shaft fitted with a brake band. Figure 42 and Colour 37.

BRAN The outer layer of the wheat berry.

BRAYER (bray.) See bridge tree.

BREAST (of a mill). The lower half of the front of the body of a post mill.

BREAST (of a stone). The middle third of the face of a millstone.

BREAST BEAM (weather beam.) The main cross-beam at the front of the mill supporting the weight of the wind shaft.

BRIDGE See rynd.

BRIDGE BOX (bridging box.) The adjustable bearing on the bridge tree that centres the spindle in neck box. Figure 43 and Colour 45.

BRIDGE TREE (brayer, bray.) The pivoted beam that supports the stone spindle so that the gap between the stones is controlled. Figure 43 and Colour 45.

BUCK See body.

BURR STONE French. A mill stone usually made of shaped pieces of quartzite which was imported from France and used for flour. Figure 45.

BUSHEL The ancient measure of all grains. It was a volume, not a weight, that equalled eight gallons, which for grain was 60 to 75lb depending on berry size and moisture content.

CANISTER (poll end.) An iron casting with two square holes that fits on the end of the wind shaft and holds the sail stocks. Figure 30 and Colour 32.

CANT POST The corner post of a smock mill.

CANVAS The cloth of a common sail.

CAP The turnable roof on a smock or tower mill. Figure 18.

CAP CIRCLE The underside of the cap that rests on the rollers to let the cap turn.

CAP SPARS (ribs.) The rafters in the cap.

CENTRING WHEELS (sheaves.) Cast-iron wheels fitted to the curb to stop the cap from moving off-centre during rotation. Figure 19 and Colour 27.

CHAIN POST The wooden posts round a post mill that are used to anchor the chains that pull the buck to face the wind and then hold the buck in that position.

CHAIN PURCHASE WHEEL (chain wheel.) A wheel with a V groove, so designed that a chain would not slip. Sometimes used to operate the striking gear.

CLAM An iron pincer used to grip and lift the runner stone.

CLAPPER The oldest form of vibrator that dragged on the runner stone to shake grain into the eye. Also used for a stick fixed to the top of the runner stone that banged the shoe bottom once each revolution to do the same.

CLASP ARM A way of holding wheels onto shafts. The wheel is made independently of the shaft with a square hole in the centre. The wheel is slid onto the shaft and wedged on to a square section on the shaft. The positioning of the wheel on the shaft is very precise, so that meshing of gears is exact. Figure 39.

COCK HEAD The dome-shaped top end of the spindle on which the runner stone rests.

COG (cant.) The teeth on the gear wheels which may be wood or cast iron.

COLLAR The large bearing under the buck of a post mill that holds it central as it is moved.

COMMON SAIL (cloth sail.) A simple latticed frame over which cloth sails were spread. Figure 24 and Colour 28.

COMPASS ARM WHEEL The older way of holding wheels, particularly the brake wheel, onto shafts. The spokes are mortised into the shaft so the position of the wheel on the shaft is not adjustable.

COMPOSITION STONES Stones made by mixing carborundum or emery with cement and letting the mixture set in a round mould.

COOMB See sack.

COUNTER WHEEL A large chain wheel on the cap used to turn gears or worm to rotate the cap into the wind.

CRACKING (stitching, feathering.) The chiselling of the fine lines on the lands of a mill stone. Figure 49.

CRANKS See bell crank.

CROTCH (crutch.) The lower forked end of the quant which engages the mace. Colour 40.

CROOK STRING The string used to change the slope of the shoe or the opening of the gate on the hopper to control the flow of grain into the eye.

CROSS A substantial iron casting, with four, five, six or eight arms for sails, that fits to the end of the wind shaft. Figure 30 and Colour 31.

CROSS TREES The horizontal beams of the trestle of a post mill to which are fixed the quarter bars that hold the main post.

CROWNTREE The main cross-beam of a post mill that supports the buck and pivots on the top of the main post.

CROWN WHEEL A gear wheel used in the horizontal plane with upward-facing cogs that resemble a crown. Colour 37.

CULLIN STONE, cullen stone. See blue stone.

CURB The top of the wall on a tower or smock mill that supports the cap and allows it to turn.

DAGGER POINT The second reefing position of a common sail where area is ½ of full sail and the shape of the sail resembles a dagger. Figure 24.

DAMSEL Originally fitted to under-driven stones on the top of the spindle, it is a shaft with three or four sides that shake the shoe so that grain feeds into the eye of the mill stone. The term is now sometimes applied to the square section of the quant that shakes the shoe. Figure 41.

DEAD CURB The oldest type of curb using no rollers where the two moving surfaces slid directly on one another lubricated with tallow or animal fat.

DOUBLE SHUTTERED, double-sided. A patent sail with shutters on both sides of the whip. Figure 26.

DRAFT The angle to a radius of the main furrows on the stone.

DRAFT CIRCLE The imaginary circle in the eye of a stone to which the main furrows are tangential either to the left or right.

DRESSING The chiselling of the furrows and stitches on the mill stones.

DRESSING OF FLOUR The cleaning and separation of the flour into two, three grades of fineness.

DRIVING SIDE, trailing edge. The broad part of the sail that provides most of the power.

DUST FLOOR The top floor of a smock or tower mill to which the grain sacks are hoisted.

ENDEN COUNTING The Dutch way of stating sail speed. Count the ends of the sail as they swish past in 1 minute – sixty ends per minute on a four-sailed mill is the same as 15rpm.

EYE The round hole in the centre of the runner stone into which the grain is fed for milling and the inner third of the millstone face.

EYE OF THE WIND The direction of the wind into which the sails are set.

FACE See land.

FAN See fantail.

FAN BRACE The rods between the fan blades.

FAN SPARS The spokes of the fantail.

FANG STAFF The brake lever in East Anglia.

FANSTAGE The wooden platform supporting and giving access to the fantail on a tower or smock mill.

FANTAIL (fan, fly, fly tackle, vane.) The small fan at the rear of the cap that keeps the main sails facing the wind. Figure 20 and Colour 6, 11 and 19.

FANTAIL CARRIAGE A cart at the end of the tail pole of a post mill whose wheels are driven by the fantail to move the sails into the wind.

FARM WINDMILL A multi-bladed annular sailed windmill for water pumping or DC power generation. Figure 55 and Colour 21.

FEATHER EDGE, LAND EDGE The low-angle slope of the furrow. Figure 50.

FEATHERING See cracking.

FELLOE The rim sections of a wooden wheel.

FINIAL The top of a Lincolnshire ogee cap, which is a round ball on a stem. Colour 11.

FIRST REEF The term for the common sail when reefing has reduced the area to ¾ of full sail. Figure 24.

FLATS The cast-iron bearings in which the shutters pivot.

FLY See fantail.

FLY POSTS The upright timbers which support the fantail.

FOOTSTEP The bearing supporting the main shaft.

FULL SAIL The cloth is spread over the complete framework of the sail. Figure 24.

FURROWS The grooves that are chiselled into the face of the mill stone. Figures 48, 49 50.

FURROW STRIPS (sticks.) Slats of wood used to mark out the position of the furrows for chiselling.

GALLERY See balcony.

GATE (spittle.) A small sliding door used to control the flow of grain from the hopper into the shoe.

GIN The large frame in which a horse walked round to turn a wheel that could drive stones.

GIRTS The beams running the length of the sides on a post mill that rest on the crown tree on which the buck is constructed.

GLUT BOX The bearing enabling the stone nut to be disengaged from the spur wheel.

GOLDEN THUMB, rule of thumb. A saying based on the use of the miller's thumb to test the fineness and warmth of the meal from the spout to control the milling process, but now more widely used to show good judgement based on practical experience.

GOVERNOR A system of rotating weights which change position in relation to the speed of the driving shaft. The movement is used to control the gap between the mill stones in response to changes in shaft speed caused by wind variation. Figure 54 and Colour 44.

GRAFT SHAFT A compound shaft of wood and iron.

GREASE WEDGE The adjustable part of the spindle bearing in the neck box.

GREAT SPUR WHEEL See spur wheel.

GRIPE East Anglian name for the brake.

GRIPE ARM East Anglian name for a clasp arm.

GRIST The grain to ground or animal feed.

GRIT STONE See Peak stone.

GUDGEON PIN A substantial iron pin used at the end of wooden shafts as a pivot. Figure 15.

GUY, guy ropes. The ropes from the end of the stay on a Mediterranean mill that help support the slender sail stocks.

GYMBAL BAR See rynd.

HACKLE PLATE The upper square plate in the neck box fitted with a leather washer that prevents dust getting down the side of the spindle.

HARPS (quarters.) The pattern of furrows chiselled into the face of the stones. Figure 48.

HEAD The upper part of the front of a post mill.

HEAD SICK The description of a post mill when the buck tilts forward.

HEAD WHEEL See brake wheel.

HEFT The wooden handle for the bills. Figure 46 and Colour 42.

HEEL The end of a sail closest to the wind shaft.

HEMLATH The outer, longitudinal, wooden bar of the sail frame.

HIGH GRINDING Milling with the stones more widely separated so that regrinding is needed.

HOPPER A box above the stones holding grain which feeds via the shoe into the stones.

HORIZONTAL WINDMILL A windmill in which the sails revolve in a horizontal plane about a vertical shaft, particularly those invented in early nineteenth century.

HORNS The parts of the main post of a post mill that straddle the cross trees. Figure 13 and

Colour 25.

HORSE A frame holding the hopper and shoe in position over the stones. Colour 43.

HUNTING COG The name given to the extra or omitted tooth on a gear wheel which avoids the same teeth meeting on every revolution and evens out wear.

HURST FRAME A stout wooden frame that holds mill stones when they are not let into the floor.

JACK The device used to check the verticality of the spindle.

JACK RING An iron ring used to lift the stone nut out of gear with the spur wheel.

JACK STAFF A lath of wood fitted on the spindle to test the verticality of it.

JIB SAIL A triangular cloth sail wound round a radial sail arm with the pointed end roped to the preceding sail. A system much used in the Mediterranean region.

JOCKEY PULLEY A small undriven pulley that can be moved to adjust the tension in a belt.

JOGGING SCREEN A shaking sieve to grade partly milled grain so that the oversize can be reground.

JOGSCRY (jumper.) A shaken inclined sieve used to clean grain or grade flour. It was the first of its type to be used.

KNAVESHIP (levy, toll, multure.) The portion of meal taken by the miller as payment for milling. Usually about $\frac{1}{16}$.

KEEP A wood or iron block to hold down the tail end of the wind shaft.

KEEP IRON (keep flange.) The flange on the inner side of the curb ring under which the centring wheels run to prevent lifting of the cap in strong winds.

KING POST See main post

LANDS The raised parts between the furrows of a mill stone. Figures 49 and 50.

LAND STRIP (stick.) A lath of wood the width of a land used to mark out lands on the face of the mill stone. Usually paired with a furrow stick.

LANTERN PINION (lantern gear.) A very old form of drive where staves between two discs act as teeth to engage pegs on the other wheel. Figure 4.

LATHS The longitudinal woods on the frame of common sails.

LEADER BOARDS (leading boards.) Longitudinal boards on the front edge of some sails to improve efficiency.

LEVY See knaveship.

LIGHTER SCREW The screw that sets the position of the bridge tree.

LISTINGS The straps fastened to the blinds of roller sails for reefing.

LIVE CURB (shot curb.) A curb fitted with rollers to help the smooth turning of the cap.

LOADING STAGE The brick or stone platform outside the door of a smock or tower mill at horse-cart height to make easier the task of handing heavy sacks onto or off the cart. Figure 17 and Colour 10.

LOW MILLING The traditional way of milling with the stones set close together so the grain is milled in one operation.

LUFF (luffing, winding.) Turning the cap or buck so the sails face the wind.

MACE An iron casting between the rynd and quant. Figure 43 and Colour 40.

MAIN POST (king post.) The central oak post on which the buck of a post mill rotates.

MAIN SHAFT (central, upright.) The shaft from the wallower to the spur wheel.

MEAL The ground grain emerging from the stones.

MEAL BIN The boxes that receive the meal from the meal spout.

MEAL FLOOR The floor where the meal bins are located. Usually the ground floor of a three or four floor mill or the fourth floor down from the top of a higher mill. Figure 17.

MEAL SPOUT The end of the meal chute that directs the meal into the bins or sacks.

METTLE, to show one's mettle (metal). A saying based on the fact that stone dressers suffered from specks of iron from the bill becoming embedded in the back of their hands. Lots of specks meant lots of experience and hence a skilled dresser. The miller would have asked dressers to 'show their mettle' as proof of their skill, and hence it evolved to become an expression of not only skill but bravery in any field.

MIDDLING The Kentish name for the sail stock.

MIDDLINGS (pollards.) The coarse bits of the floury part of the berry with some bran.

MILLERS WILLOW The wooden spring that holds the shoe in contact with the damsel or quant.

MILL SOKE (soke.) The ancient law restricting the building of mills to the Church or King and requiring all folk to take their corn to their Lord's mil.

MILL STONE One of a pair of round stones that grind the grain.

MILL STONE GRIT The coarse quartz-like rock beds from which millstones are quarried. The major source is in the Peak District of Derbyshire.

MORTISE JOINT The wood joint using a hole cut through one timber with a second timber wedged into it at right angles. Before cast iron, the way of fixing the sail stocks into the wind shaft.

MULTURE See knaveship.

MULTI-BLADED See annular sails.

NECK BEAM The front cross-beam that supports the front end of the wind shaft.

NECK BEARING The front bearing of the wind shaft that takes most of the weight. Figure 38 and Colour 35.

NECK BOX The bearing in the bed stone for the spindle on which the runner stone revolves.

NECK OF WIND SHAFT The front part of the wind shaft.

OGEE CAP (onion.) The onion-shaped cap typical of Lincolnshire. Figure 7 and Colour 5,6 and 11.

OVERDRIFT (overdrive.) Stones that are driven from above. Figure 40.

PADDLE The plate fixed to the runner stone to push the meal into the meal chute.

PAINT STAFT A straight length of wood daubed on the underside with red paint to indicate high spots on the face of a mill stone.

PANNIER An extension on the side wall of a post mill. Figure 16.

PATENT SAIL The weight-controlled shuttered sails invented and patented by Cubitt in 1807. Figure 25 and Colour 29 and 30.

PEAK STONE (grit, Derbyshire, grey.) A one-piece mill stone quarried from the Peak District of Derbyshire.

PEG MILL The earliest form of post mill built on a post buried deep in the ground without side supports.

PETTICOAT The short boards round the edge of the cap or buck to deflect rain away from the tower or trestle.

PIERS (pillars.) The brick or stone plinths that support the ends of the cross trees of a post mill. Figures 14 and 16.

PICK See bill.

PINCH BAR The lever used for the initial lifting of the runner stone so that wedges can be inserted and removal completed.

PINION The smaller gear wheel of a pair. Colour 27.

PINTLE The pivot at the top of the main post of a post mill. Figure 15.

PITCH The angle of the sail from right angles to the wind.

POINTING LINES The ropes fixed to the cloth of common sails for reefing.

POLLARDS See middlings.

POLL END See canister.

PORCH An extension on the back wall of a post mill. Figure 16.

POST MILL A wooden mill in which the whole body containing stones, storage and sails, turns on a central massive oak post. Figure 5 and Colour 1 and 2.

PRICK POST The central upright in the breast of a post mill.

PRITCHELL A stone chisel used with a hammer to dress the stones – a rare alternative to the bill.

PROOF STAFF An iron bar which is truly flat and straight to check the accuracy of the paint staff before use. It would be kept in a substantial wooden box for protection.

PUNCHEONS The horizontal beams bracing the cap circle.

QUANT, crotch pole, crutch pole. The square shaft from stone nut to the mace. Figure 38 and 40 and Colour 40.

QUARTER A measure of grain equal to eight bushels or 500 to 600lb. It may refer to a quarter of a horse load of about one ton.

QUARTERS See harps.

QUARTER BARS The wooden beams from the crosstrees to the central post of the post mill. Inclined at about 45 degrees, they transfer the full load of the buck from the post to the outer part of the crosstrees. Figures 13 and 14 and Colour 1 and 25.

QUARTER DRESS The pattern on a mill stone with straight furrows in sectors called harps or quarters – even though there are seldom only four sectors. Seven to eleven is more usual.

QUARTERING Turning the sails so they are at right angles to the wind.

QUARTER (of a smock mill). One of the sides of the mill.

QUERN A pair of small-diameter stones for hand grinding. Figure 1.

RACK The line of cogs round the curb of a smock or tower mill. Colour 27.

RADDLE (tiver.) The mixture of red iron oxide with tallow, fat or water that is used on the paint staff to find the high spots on the mill stones when dressing.

RAP The block of hard wood on the shoe that takes the knock of the damsel or quant.

RED OXIDE Iron oxide used with tallow, fat or water to coat the paint staff.

RED STONE A mill stone used in the north-west of England.

REEFING The reduction of sail area by rolling up and tying cloth sails.

RIGGER Chain or strap used to disengage the stone nut from the spur wheel – an alternative to the jack ring.

RODE BALK See breast beam.

ROLLER REEFING SAIL A method of control of sail area as wind speed changes. The cloth sail is made to furl or unfurl automatically as wind speed changes. Figures 31, 32, 33 and 34.

ROUNDHOUSE The brick or stone building at the base of a post mill whose roof fits beneath the buck to protect the trestle from rain and provide storage space. Figures 5 and 16, Colour 2.

RUBBING BURR A hand-sized piece of burr stone that can be used to remove high spots when dressing stones.

RULE OF THUMB See golden thumb.

RUNNER STONE The upper driven stone of a pair of mill stones. Figures 47 and 51.

RYND (rind, rhynd, bridge, bridge iron, gymbal bar.) The iron bar, usually leaded into the top runner stone, by which the runner stone is located on its spindle. The most common is a simple bar but three and four-arm versions are used. Figure 41.

SACK, coomb. An old measure of grain equal to four bushels.

SACK BOY A wooden slat with hooks to hold the mouth of a sack open.

SACK CHAIN (sack hoist.) The chain, often endless, used to raise grain sacks to the dust or bin floor.

SACK SLIDE A plank fixed to the side of post mill steps to get the meal sacks down.

SAILS Sometimes called sweeps or wands.

SAIL BACK See whip.

SAIL BARS The short cross-bars of a sail.

SAIL IN Close the shutters.

SAIL OUT Open the shutters.

SAIL RODS See shutter bar.

SAMSON HEAD The iron castings on the top of the post and underside of the crown tree that support the pintle.

SCOTCH WEDGE The stepped wedge used to raise the runner stone.

SHEAVE OF CORN A bundle of cut wheat stalks bound with twisted stalks about 30cm in diameter that can be stacked in ricks to help drying.

SHEAVES See centring wheels.

SHEER TREES (sheers.) The two long beams from breast to tail of a post mill that support the buck or the two main beams in the cap of a tower or smock mill.

SHOE The inclined chute from the hopper that directs the grain into the eye of the stone. Figure 51 and Colour 43.

SHOT CURB See live curb.

SHUTTER BAR See sail rod.

SHUTTERS (shades, vanes.) The moveable small sails made of canvas on a frame or board that are fixed in bays at right angles to the sail stock. Their position is controlled by weights or springs to change the opening of them as wind speed strengthens. Figure 25 and Colour 9, 11, 29 and 30.

SICKLE DRESS (Dutch.) A stone with curved furrows from eye to rim. Figure 48.

SINGLE SHUTTERED Sails with shutters on the trailing side only. Colour 9 and 10.

SKIRT The outer third of the surface of a mill stone.

SKY-SCRAPERS The name used in Suffolk for the longitudinal boards on the leading edge of sails.

SLIP COGS The removable section of cogs on a gear.

SLIPPER See shoe.

SMOCK MILL A wooden-framed and sheeted tower mill. Figures 6 and 17 and Colour 3 and 4.

SMUT (smutter.) The mould on grain and the spinning machine that removes it.

SOKE See mill soke.

SPATTLE See gate.

SPIDER The spoked device at the end of the striking rod on patent sails that operates the cranks that control the position of the shutters. Figures 35 and 36 and Colour 33.

SPILL THE WIND The opening of the shutters to control speed or avoid over-speed in gusts.

SPINDLE A small-diameter iron shaft on which other wheels rotate.

SPINDLE BEAM The beam in a post mill that holds the bearing of the central shaft.

SPRATTLE BEAM The fixed horizontal beam just below the cap that holds the upper bearing of the central shaft in a tower or smock mill.

SPRING SAILS Shuttered sails operated by springs. They were the earliest form of shuttered sail invented by Meikle. Figure 25.

SPUR WHEEL (great spur wheel.) The large gear wheel rotating horizontally from which several other small pinions are driven at an increased speed. Figures 40 and 41 and Colour 38.

STAFF See paint staff.

STAGE The platform on the fantail gantry or the balcony.

STAR WHEEL The iron hub casting into which fit the spars for the fantail vanes.

STAVES The wooden rods of a lantern pinion that act like cogs on a pinion.

STAY On Mediterranean mills, the smaller diameter extension on the wind shaft, to the end of which are tied ropes that go to the end of the slender sail stocks to help support them.

STEELYARD The long iron lever between the bridge tree and the governor.

STITCHING See cracking.

STOCKS (sail stocks.) These are the main beams that carry the sails. They are fixed to the wind shaft and on older smaller mills the sail bars are mortised directly into them. On larger mills, the stocks carry a second beam called a whip to extend sail length.

STONE CASING See vat.

STONE FLOOR The floor in the mill where the stones are. Figure 17.

STONE NUT The pinion that engages the spur wheel and drives the runner stone. Figures 40 and 41 and Colour 38.

STONE SPINDLE The iron shaft on which the runner stone is balanced and rotates. Figures 51 and 54 and Colour 45.

STORM HATCH A removable shutter in the cap above the wind shaft that gives access to poll end and sails.

STRIKE A volume measure equal to a third of a bushel.

STRIKING GEAR The mechanical linkage that closes and opens shutters on a patent sail or furls and unfurls roller-reefing sails. Figures 35 and 36 and Colour 33.

STRIKING ROD The iron shaft down the centre of the wind shaft that operates the striking gear.

STUMP IRONS The supporting frame on the stocks for the cranks of the striking gear.

SUBSTRUCTURE See trestle.

SWALLOW The increased gap between the stones at the eye to admit grains for grinding.

SWEEP A southern name for a sail.

SWORD POINT Description of common sails when reefed to minimum area of about ¼ full sail, so-called because it looks like the tapered end of a sword blade. Figure 24.

TAIL BEAM The beam holding the tail bearing of the wind shaft.

TAIL POLE The large beam fitted under and at the back of the post mill body used to turn the body and sails into the wind. Figure 14.

TAIL WHEEL In post mills, the second gear wheel fitted to the back or tail end of the wind shaft to drive a second pair of stones.

TAIL WINDING A dangerous situation when the wind direction has changed suddenly and is blowing on the back of the sails.

TALTHUR A lever pivoted on the side of the tail pole to raise the steps so the buck can be turned. Figures 5 and 14.

TENTERING GEAR The system of levers used to alter the gap between the stones.

TENTERING SCREW The screw the miller can use to adjust the stone gap. It operates in parallel with the governor. Figure 54.

THRIFT The ash or beech handle into which the bills are wedged.

THRUST BLOCK The block on the tail beam that holds the thrust bearing of the wind shaft.

TIVER See raddle.

TOLL See knaveship.

TOWER MILL A windmill using a brick or stone slightly conical tower that contains the milling machinery and a rotating cap so that the sails can be manoeuvred into the wind. Figure 17 and Colour 5 to 19.

TRAMMEL A wooden frame used to check the true vertical of the stone spindle.

TRAMMING The use of the trammel to true the spindle.

TRESTLE The wooden structure that supports the body of a post mill and allows it to turn. Figure 5, 13 and 14 and Colour 21.

TRIANGLE See bell cranks.

TRIPOD MILL An early form of post mill which had three bracing bars on the main post.

TRUCK WHEELS See centring wheels.

TRUNDLE WHEEL A wooden gear wheel having pegs projecting from the face instead of cogs.

TUN See vat.

TWIST PEG A wooden peg near the meal spout to which the cord controlling grain flow into the stones is tied.

UNDERDRIFT Millstones driven from below. Figure 41.

UNDER RUNNER The lower stone when driven with upper stone stationary.

UPLONG The longitudinal bars of the sail frames of common sails.

UPRIGHT SHAFT See main shaft.

VANE See fantail and shutters.

VAT The circular or many-sided removable box round the millstones. Figure 51 and Colour 43.

VERTICAL WINDMILL The oldest type with an upright shaft directly driving a single pair of stones.

WALLOWER (crown wheel.) The horizontal gear wheel driven by the brake wheel. Figures 39 and 40 and Colour 37.

WEATHER (weather angle.) The twist or gradual change in angle of the sail along its length that improves aerodynamic efficiency. Colour 30.

WEATHER BEAM See breast beam.

WHIP The sail beam fixed to the sail stock, cross or canister to increase the length of the sail. Figure 24 and Colour 8.

WIND FARM The description when several wind turbines are built in one place.

WINDING See luffing.

WINDSHAFT The principal shaft of the windmill. It carries the sails that turn it and the brake wheel (and tail wheel in a post mill) to get the power to the stones. Figures 17 and 40.

WIND TURBINE The description used for modern, usually very large, electric power generators. Figures 58 and 59 and Colour 22, 23 and 24.

WINNOWER A fan which blows dust from the grain as it falls down a chute.

WIRE MACHINE A dressing machine that brushes the meal through sieves into several grades.

WORM A big spiral thread on a shaft that engages the rack to slowly turn the cap.

YOKE The vertical wooden bars on the tail pole used by the miller to push the buck of a post mill into the wind.

Y-WHEEL A wheel with split iron bars opened to form a Y round the outer rim that carry the sack hoist chain. Colour 41.

WINDMILL DATA SHEET

Name _____ Built _____

LOCATION _____

TOWER or SMOCK MILL

Height _____
Floors _____
Cap type _____
Curb diam _____
Base diam _____

POST MILL

Height _____
Floors _____
Roof Type _____
Panniers, Porch _____
Trestle _____
Roundhouse _____

LUFFING

Fantail or tail pole _____
Fantail blades _____
Fantail diam _____
Ratio, fantail to cap _____

SAILS

Type – patent, spring or common _____
No. _____
Fixing to wind shaft-canister or cross _____
Design: single, double sided, wind board _____
Arm divide _____
Arm length _____
Sail Length _____
Sail Width _____
Rotation _____
Bays/sail _____
Shades/bay _____

Angle to wind, heel _____
               tip _____
Shaft angle _____

Swept area _____

Total area _____
As % of swept area _____
Tip speed at 15rpm sail _____
TSR in 15mph wind _____

WHEELS	Type	Made of	Diameter	No. of teeth
Brake				
Wallower				
Spur				
Stone nut				
Stone to sails ratio				
Brake type				

STONES

Type _____
No. _____
Diameter _____
Rotation _____
Dressing, furrows, straight, curved _____
    Harps, No. _____
    Furrows per harp _____
    Geometry – radial, tangential left or right _____
Governor _____
Speed at 15rpm sail _____
Rim speed _____
Sail to stone area _____
Production rate _____

Unusual features
_____
_____
_____
_____
_____
_____
_____
_____
_____
_____
_____
_____

Literature in Date Order

1674	'Windmill for Raising Water from Mines.' Johnson, Patent No.174
1684	'Sails for Windmills.' Heckford, Patent No.243
1738	'Engine for Raising Water.' Kay, Patent No.561
1744	'Machine for Grinding Corn.' Perkins, Patent No.609
1745	'Self-regulating wind machine.' Lee, Patent No.896
1749	'Wind Hoisting Machine.' Langworthy, Patent No.643
1759	'An Experimental Enquiry concerning the Natural Powers of Water and Wind to turn Mills, and other Machines, depending on a Circular motion.' Smeaton. *Philosophical Transactions*, Vol.51 (1759–1760)
1768	'Machine for Dressing and Cleaning Grain.' Meikle and Mackel, Patent No.896
1783	'Sails for Windmills.' Wiseman, Patent No.1,399
1785	'Windmills.' Hilton, Patent No.1,484
1787	'Regulator for Wind and other Mills.' Mead, Patent No.1,628
	'Regulating the Sails of Mills.' Heame, Patent No.1,588
1788	'Machine for Separating Corn from Straw.' Meikle, Patent No.1,645
1789	'Regulating Wind and other Mills.' Hooper, Patent No.1,706
1792	'Wind and Watermills.' Silvester, Patent No.1,890
1796	*On Mills*, Telford, copy of his manuscript at Institute of Civil Engineers
1804	'Windmills.' Bywater, Patent No.2,782
1807	'Windmills.' Cubitt, Patent No.3,041
1861	*Mills and Millwork*, two vols, Sir William Fairburn. Longmans Green.
1868	'Windmills.' Warner and Chopping, Patent No.938
1873	'Windmills.' Hammond, Patent No.1,654
	Spons' Dictionary of Engineering
1885	*The Windmill as a Prime Mover*, Wolfe, John Willey, 159pp, 40 illustrations
1887	*Special Report of British & Irish Millers Association into the Depressed State of the Milling Industry*
1898	*History of Corn Milling*, Bennet and Elton, Simpkin Marshall

Vol.1 *Hand stones, slave and cattle mills*

Vol.2 *Water and windmills*

Vol.3 *Feudal law and customs of mills*

Vol.4 *Some famous feudal mills*

1910   *Standard Cyclopedia of Modern Agriculture*, Vols 1–10

1916   *Windmill Land*, Clarke, Dent

         A Picturesque History of Yorkshire, Vols 1–6

1923   *Windmills*, Brangwyn, Frank and Preston

1930   *English Windmills*, Vol.1, Batten Architectural Press

1931   *In Search of English Windmills*, Hopkins and Freese, Cecil Palmer

         England of the Windmills, Mais, Dent and Sons

1932   *Life in Rural England*, W.C. Finch, C.W. Daniel & Co.

         English Windmills, Vol.2, Smith, Architectural Press

1936   *Windmills in Sussex*, Hemming, Daniel

         On Mills, Telford, Edited by Lancaster Burne., Trans of the Newcomen Soc., 1936–37, Vol.17

1945   *Through the Mill*, Burnett, Epworth Press

1948   *Power from the Wind*, Putnam, Vann Strand

1951   *Lincolnshire Windmills*, Wails, Trans of the Newcomen Soc.

         Vol.28 *Post mills*

         Vol.29 *Tower mills*

1952   *Flour for Mans Bread*, Storck and Teague, University of Minnesota

1954   *The English Windmill*, Wails, Routledge & Kegan Paul

1955   *The Making of the English Landscape*, W.G. Hoskins, Hodder & Stoughton

1956   *Watermills and Windmills of Middlesex*, Blythman, Baron

1960   *Windmills of Hampshire*, Triggs, Down Memory Lane

1961   *Windmills of Derbyshire, Leicestershire, & Notts – Post Mills*, Baber & Wailes

1962   *The Dutch Windmill*, Stokhuyzen, Van Dishoeck

1963   *Cornish Windmills*, Douch, Oscar Blackford

1968   *Discovering Windmills*, Vince, Shire Publications

         Age of Industrial Expansion, Holland, Nelson

1970   *Windmills and Watermills*, J. Reynolds, Hugh Evelyn

         Windmills, Rise and Decline, Wailes

         Mills of the Isle of Wight, Major, Charles, Skilton

         Windmills of Norfolk, Smith, Stevenage Museum Publications

1971   *Windmills and Millwrighting*, Freese, David & Charles

1972   *The Windmill*, De Little, John Baker

         Windmills of Huntingdon and Peterborough, Smith, Stevenage Museum Publications

1973   *Wind Engines (Farm), A Necessary Study*, Major

         Windmills in Bucks and Oxon, Smith, Stevenage Museum Publications

         Windmills of Kent, West, Charles Skilton

1975   *Windmills in England*, Wails, Charles Skilton

         Windmills, Beedle, Bracken Books

	Windmills in Sussex, Smith, Stevenage Museum Publications
	Windmills in Cambridgeshire, Smith, Stevenage Museum Publications
	Windmills in Bedfordshire, Smith, Stevenage Museum Publications
1976	*Windmills of England*, Brown, Robert Hale.
	Wind Catchers, Torrey, Stephen Green Press, Vermont
	Windmills of Surrey and London, Smith, Stevenage Museum Publications
1977	*Old Watermills and Windmills*, Hopkins, E.P. Publishing
	The Windmills of John Wallis Titt, Major, International Molinological Society Symposium
	The Restoration of Windmills and Water pumps in Norfolk, Scott, Norfolk Windmills Trust
	Windmills in Warwickshire, Smith, Stevenage Museum Publications
1978	*The Lost Windmills of Retford*, Briggs, Eaton Hall
	Windmills in Suffolk, Smith, Stevenage Museum Publications
1979	*Windmills and Watermills*, Wails, Ward Lock
	Englands Vanishing Windmills, A.E.P. Shillingford, Godfrey Cave Associates
	Windmills of Sussex, Brunnarius, Phillimore
	Windmills in Staffordshire, Seaby and Smith, Stevenage Museum Publications
1980	*Wind Energy*, Kovari, Domus Books
	Growth of the British Economy 1700–1850, Speed, Wheaton
1983 –6	'Corn Mill Sites in Cheshire', Bott, *Cheshire History,* Vols 10, 11, 14, 15, 16, 17
	Windmills of Shropshire, Hereford & Worcester, Seaby and Smith, Stevenage Museum Publications
	Bedfordshire Mills, Howes, Bedford County Council
1984	*Scottish Windmills – A Survey*, Douglas, Oglethorpe and Hulme, Scottish Industrial Archaeology Survey
1985	*East Yorkshire Windmills*, Gregory, Skilton
1986	*Lincolnshire Windmills – A Survey*, Dolman, Linconshire County Council
	Windmills in Hertfordshire, Smith, Stevenage Museum Publications
1987	*Harvesting the Air, Windmill Pioneers in the Twelfth Century*, Kealey, Boydel Press, 307pp, 25 illustrations
1988	*The Mills of Medieval England*, Holt, Basil Blackwell, 202pp, 14 illustrations
1990	*Reflections of a Bygone Age: Lincolnshire Windmills on Old Postcards*, Croft
	Drainage Mills of Norfolk, Smith, Stevenage Museum Publications
1991	*Yorkshire Windmills*, Whitworth, M.T.D. Rigg Publications
1991	*Windmills of Northamptonshire*, T.L. Stainwright, Wharton
1993	*Windmills and How They Work*, Vince, Sorbus
1994	*Power from the Wind*, Hills, Cambridge University Press
1998	*Hewitt's Windmill*, Hewitt, Published by Hewitt
1999	*Hertfordshire Windmills and Wind Millers*, C. Moore, Windup Publications
2000	*Wind Energy*, Riddle, E.W.E.A. Summary
2001	*Tyke Towers, Yorkshire Windmills*, Whitworth, Landy
2004	*The Solar Economy*, Sheer, Earthscan

Mills Open, Woodward-Nut, Society For the Protection of Ancient Buildings
Wind Power, Gipe, James & James
2005   B.W.E.A. Briefing Sheets
The Industrial Windmill in Britain, Gregory, Philmore
Windmills, A Pictorial History of their Technology, Hills, Landmark Publications

Index

Visit our website and discover thousands of other History Press books.

www.thehistorypress.co.uk